Universitext

Günter Tamme

Introduction to Étale Cohomology

Translated by Manfred Kolster

Springer-Verlag Berlin Heidelberg GmbH

Günter Tamme

Mathematisches Institut, Universität Regensburg
Universitätsstrasse 31, D-93053 Regensburg, Germany

Translator:

Manfred Kolster

Department of Mathematics and Statistics, McMaster University
Burke Science Building 132, 1280 Main Street West
Hamilton, Ontario, Canada L8S 4K1

Mathematics Subject Classification (1991):
14-01, 11-01, 11Gxx, 18Gxx, 14F20

ISBN 978-3-540-57116-2 ISBN 978-3-642-78421-7 (eBook)
DOI 10.1007/978-3-642-78421-7

Library of Congress Cataloging-in-Publication Data. Tamme, Günter, 1937- . Introduction to Étale cohomology / Günter Tamme; translated by Manfred Kolster. p. cm. – (Universitext) Includes bibliographical references and index. ISBN 978-3-540-57116-2
1. Geometry, Algebraic. 2. Homology theory. 3. Sheaf theory. I. Title.
QA564.T36 1994 514'.23–dc20 94-3558

Typesetting: Camera ready by author
SPIN: 10124343 41/3140 – 5 4 3 2 1 0 – Printed on acid-free paper

Preface

The present book has its origin in a lecture course, which I gave at the University of Göttingen in 1975/76. The manuscript that emerged from these lectures was reproduced in the series "Der Regensburger Trichter" by the Department of Mathematics at the University of Regensburg.

This introductory book presents the basic concepts and methods of étale cohomology. After reading this book, the reader should be able to understand without difficulty the more advanced literature such as SGA 4 by A. Grothendieck, M. Artin, J. L. Verdier, SGA $4\frac{1}{2}$ by P. Deligne and Étale Cohomology by J. S. Milne.

Étale Cohomology uses the language of homological algebra in abelian categories. Therefore I have summarized in a preliminary chapter the relevant facts, as, for example, right derivatives of functors and spectral sequences.

Chapter I deals with Grothendieck Topologies; in particular, the étale topology of a scheme is a Grothendieck Topology. The general consideration of this topology is necessary, for instance, in order to compare different cohomologies via Leray spectral sequences.

Chapter II deals exclusively with étale cohomology of schemes. In §10 we shall take a closer look at the cohomology of 1-dimensional noetherian schemes, which are of great interest especially for number theory. Finally, we give an overview of general theorems in étale cohomology of higher dimensional schemes.

The English edition of this book was made possible by my friend and colleague Jürgen Neukirch. He arranged the translation, negotiated with the publishers, shaped the outward appearance of the manuscript and made it ready for print. For this I would like to thank him. Finally, I would like to express my gratitude to Manfred Kolster for his efforts translating this book.

Regensburg, March 1994 Günter Tamme

Contents

Chapter 0
Preliminaries

§ 1. Abelian Categories

(1.1) Categories and Functors

Let $\mathcal{C}$ be a category and let $u : A \to B$ be a morphism in $\mathcal{C}$. Then u is called a **monomorphism** (or **injective**) if the map $\operatorname{Hom}(C, A) \to \operatorname{Hom}(C, B)$ which sends v to uv is injective for all objects C in $\mathcal{C}$. Analogously, u is called an **epimorphism** (or **surjective**) if the map $\operatorname{Hom}(B, C) \to \operatorname{Hom}(A, C)$ which sends w to wu is injective for all $C \in \mathcal{C}$. The morphism u is **bijective** if u is both injective and surjective. An isomorphism, i.e. a morphism having an inverse, is always bijective. The converse is not true in general.

Let $u : B \to A$ and $u' : B' \to A'$ be monomorphisms. Then u' **dominates** u – written as $u' \geq u$ – if u factors as $u'v$ for some morphism $v : B \to B'$, which then is uniquely determined. The monomorphisms u and u' are called **equivalent**, if both $u' \geq u$ and $u \geq u'$. The corresponding morphisms $B \to B'$ and $B' \to B$ are then mutually inverse. Choose a representative from each class of equivalent monomorphisms with values in A. These representatives are called **subobjects** of A. Hence a subobject of A is an object B together with a monomorphism $u : B \to A$, which is called the **canonical injection** of B into A.

In a similar manner we can consider, for a given object A, classes of equivalent epimorphisms $A \to B$ and define the **quotients** of A as a complete system of representatives of these classes.

Let $(A_i)_{i \in I}$ be a family of objects in $\mathcal{C}$. An object A with morphisms $u_i : A \to A_i$ is called the **product of the family** $(A_i)_{i \in I}$ if the natural map

$$\operatorname{Hom}(B, A) \to \prod_{i \in I} \operatorname{Hom}(B, A_i)$$

is bijective for each $B \in \mathcal{C}$. The object A is often written as $\prod_{i \in I} A_i$, and u_i is called the i-th projection of the product.

Similarly, given a family $(A_i)_{i \in I}$ of objects in $\mathcal{C}$, an object A with morphisms $v_i : A_i \to A$ is called the **direct sum** of this family if the natural map

$$\mathrm{Hom}(A, B) \to \prod_{i \in I} \mathrm{Hom}(A_i, B)$$

is bijective for each $B \in \mathcal{C}$. The object A is written as $\coprod_{i \in I} A_i$ or $\bigoplus_{i \in I} A_i$, and v_i is called the i-th injection of the direct sum.

Let $\mathcal{C}$ and $\mathcal{C}'$ be categories, and let F and G be two covariant functors from $\mathcal{C}$ to $\mathcal{C}'$. A **morphism of functors** (or a **natural transformation**) f from F in G assigns to each object A in $\mathcal{C}$ a morphism $f(A) : F(A) \to G(A)$ in $\mathcal{C}'$ so that, for each morphism $u : A \to B$, the diagram

$$\begin{array}{ccc} F(A) & \xrightarrow{f(A)} & G(A) \\ {\scriptstyle F(u)}\downarrow & & \downarrow{\scriptstyle G(u)} \\ F(B) & \xrightarrow{f(B)} & G(B) \end{array}$$

commutes. f is called an **isomorphism** if $f(A)$ is an isomorphism for each A in $\mathcal{C}$.

A functor $F : \mathcal{C} \to \mathcal{C}'$ is called an **equivalence of categories** if there exists a functor $G : \mathcal{C}' \to \mathcal{C}$ **quasi-inverse** to F, i.e. $G \circ F$ is isomorphic to the identical functor $id_{\mathcal{C}}$ of $\mathcal{C}$ and $F \circ G$ is isomorphic to $id_{\mathcal{C}'}$.

One can show that $F : \mathcal{C} \to \mathcal{C}'$ is an equivalence of categories if and only if F is **fully faithful** (i.e. for each pair A, B of objects in $\mathcal{C}$ the map $u \to F(u)$ from $\mathrm{Hom}(A, B)$ to $\mathrm{Hom}(F(A), F(B))$ is bijective) and each object A' in $\mathcal{C}'$ is isomorphic to an object of the form $F(A)$.

Given a functor $F : \mathcal{C} \to \mathcal{C}'$, a functor $G : \mathcal{C}' \to \mathcal{C}$ is called **left adjoint** to F if there exists for each pair of objects $A' \in \mathcal{C}'$ and $B \in \mathcal{C}$ an isomorphism

$$\mathrm{Hom}(A', F(B)) \cong \mathrm{Hom}(G(A'), B),$$

which is functorial in A' and B. The functor G is then determined by F up to a unique isomorphism and is denoted by ${}^{ad}F$. The condition defining the adjoint induces canonical morphisms of functors

$$\rho : id_{\mathcal{C}'} \to F \circ {}^{ad}F$$
$$\sigma : {}^{ad}F \circ F \to id_{\mathcal{C}}$$

the so-called **adjoint morphisms**. In order to prove the existence of the functor left adjoint to a given functor $F : \mathcal{C} \to \mathcal{C}'$, it suffices to check the following: For each A' in $\mathcal{C}'$ the covariant functor

$$B \mapsto \mathrm{Hom}(A', F(B))$$

from $\mathcal{C}$ to the category of sets is **representable**, i.e. there exists an object ${}^{ad}F(A') \in \mathcal{C}$ and for each $B \in \mathcal{C}$ an isomorphism

$$\mathrm{Hom}(A', F(B)) \cong \mathrm{Hom}({}^{ad}F(A'), B)$$

which is functorial in B. From this, we obtain first of all for each $A' \in \mathcal{C}'$ a canonical morphism $\rho(A') : A' \to F({}^{ad}F(A'))$. Now, given a morphism $u' : A' \to A'_1$ in $\mathcal{C}'$, we get the morphism

$$A' \xrightarrow{u'} A'_1 \xrightarrow{\rho(A'_1)} F({}^{ad}F(A'_1))$$

in $\mathcal{C}'$, hence finally the morphism

$${}^{ad}F(u') : {}^{ad}F(A') \to {}^{ad}F(A'_1)$$

by using the isomorphism from above. Therefore ${}^{ad}F$ becomes a covariant functor $\mathcal{C}' \to \mathcal{C}$ in such a way that the isomorphisms

$$\mathrm{Hom}(A', F(B)) \cong \mathrm{Hom}({}^{ad}F(A'), B)$$

are also functorial in A'.

(1.2) Additive Categories

A category $\mathcal{C}$ is **additive** if the following properties hold:

(a) For each pair of objects A, B in $\mathcal{C}$ the set $\mathrm{Hom}(A, B)$ has the structure of an abelian group such that the composition of morphisms is bilinear.

(b) The product and sum of two objects exist in $\mathcal{C}$.

(c) $\mathcal{C}$ has a zero object, i.e. an object both final and initial in $\mathcal{C}$. Any two zero objects in $\mathcal{C}$ are canonically isomorphic.

All zero objects are then identified and the single zero object is denoted by O.

Let $\mathcal{C}$ be an additive category, and let $u : A \to B$ be a morphism in $\mathcal{C}$. We consider all monomorphisms $i : A' \to A$ with the following property: For a given morphism $C \to A$ the composite $C \to A \xrightarrow{u} B$ is the zero morphism if and only if $C \to A$ factors as $C \to A' \xrightarrow{i} A$. All these monomorphisms $i : A' \to A$, if there are any, form a full equivalence class of monomorphisms (cf. (1.1)), and hence there is precisely one which represents a subobject of A. It is called the **kernel of** u and is denoted by $\ker u$. In a similar manner the **cokernel of** u, if it exists, is defined as a quotient of B and denoted by $\operatorname{coker}(u)$.

Finally, we define the **image** and the **coimage of** u as $\operatorname{im}(u) = \ker(\operatorname{coker}(u))$ and $\operatorname{coim}(u) = \operatorname{coker}(\ker(u))$. If u has an image and a coimage, there exists a uniquely determined morphism

$$\bar{u} : \operatorname{coim}(u) \to \operatorname{im}(u),$$

such that the composition

$$A \to \operatorname{coim}(u) \xrightarrow{\bar{u}} \operatorname{im}(u) \to B$$

with the canonical surjection $A \to \operatorname{coim}(u)$ and the canonical injection $\operatorname{im}(u) \to B$ is equal to the morphism u. To see this, first note that $A \to B$ factors as $A \to \operatorname{im}(u) \to B$, since the composite $A \to B \to \operatorname{coker}(u)$ vanishes. Furthermore, $\ker(u) \to A \to B$ vanishes and therefore also $\ker(u) \to A \to \operatorname{im}(u)$, since $\operatorname{im}(u) \to B$ is injective, so that finally $A \to \operatorname{im}(u)$ factors as $A \to \operatorname{coim}(u) \to \operatorname{im}(u)$.

A functor F from an additive category $\mathcal{C}$ into another additive category $\mathcal{C}'$ is called **additive** if $F(u+v) = F(u) + F(v)$ holds for all morphisms $u, v : A \to B$ in $\mathcal{C}$.

(1.3) Abelian Categories

An **abelian category** is an additive category $\mathcal{C}$ with the following two properties:

- **Ab 1)** Each morphism in $\mathcal{C}$ has a kernel and a cokernel.
- **Ab 2)** For each morphism u in $\mathcal{C}$ the canonical morphism $\bar{u} : \operatorname{coim}(u) \to \operatorname{im}(u)$ is an isomorphism.

In particular, it follows that in an abelian category each bijective morphism (cf. (1.1)) is an isomorphism.

Moreover, if A is an object in the abelian category $\mathcal{C}$ there is a natural bijection between the class of subobjects of A and the class of quotients of A, which assigns to each subobject B of A its cokernel – denoted by A/B – and to each quotient of A its kernel.

A sequence $A \xrightarrow{u} B \xrightarrow{v} C$ of morphisms in the abelian category $\mathcal{C}$ is called **exact** if $\ker(v) = \mathrm{im}(u)$. A sequence $0 \to A \to B \to C \to 0$ in $\mathcal{C}$ is exact if and only if the sequence

$$0 \to \mathrm{Hom}(X,A) \to \mathrm{Hom}(X,B) \to \mathrm{Hom}(X,C)$$

of abelian groups is exact for all $X \in \mathcal{C}$.

A covariant functor F from the abelian category $\mathcal{C}$ into an abelian category $\mathcal{C}'$ is called **left exact** (resp. **right exact**) if for each exact sequence $0 \to A' \to A \to A'' \to 0$ in $\mathcal{C}$ the sequence $0 \to F(A') \to F(A) \to F(A'')$ (resp. the sequence $F(A') \to F(A) \to F(A'') \to 0$) is exact in $\mathcal{C}'$. A functor which is both left and right exact is called **exact**.

The most prominent example of an abelian category is the category Ab of abelian groups. Other examples are obtained from the following:

(1.3.1) Proposition. *Let $\mathcal{C}$ and $\mathcal{C}'$ be categories with $\mathcal{C}'$ abelian. Then the category $\mathcal{H}om(\mathcal{C},\mathcal{C}')$ of functors from $\mathcal{C}$ to $\mathcal{C}'$ (whose morphisms are the natural transformations of functors) is an abelian category.*

Moreover, the sequence $F' \to F \to F''$ is exact in $\mathcal{H}om(\mathcal{C},\mathcal{C}')$ if and only if for all objects $A \in \mathcal{C}$ the sequence $F'(A) \to F(A) \to F''(A)$ is exact in $\mathcal{C}'$.

Proof: a) Given $F, G \in \mathcal{H}om(\mathcal{C},\mathcal{C}')$ and natural transformations $f, g : F \to G$, the sum $f+g$ is defined by $(f+g)(A) = f(A) + g(A)$ for A in $\mathcal{C}$. This makes $\mathrm{Hom}(F,G)$ an abelian group in such a way that the composition of morphisms is bilinear.
b) The functor $0(A) = 0$ for all $A \in \mathcal{C}$ is the zero object in $\mathcal{H}om(\mathcal{C},\mathcal{C}')$.
c) Given $F_1, F_2 \in \mathcal{H}om(\mathcal{C},\mathcal{C}')$ we define a functor $F_1 \oplus F_2$ from $\mathcal{C}$ to $\mathcal{C}'$ via $(F_1 \oplus F_2)(A) = F_1(A) \oplus F_2(A)$ for A in $\mathcal{C}$. Together with the morphisms $F_i \to F_1 \oplus F_2$, defined by $F_i(A) \to F_1(A) \oplus F_2(A)$, the functor $F_1 \oplus F_2$ is the direct sum in the category $\mathcal{H}om(\mathcal{C},\mathcal{C}')$. The product is constructed along the same lines.
d) Let $f : F \to G$ be a morphism in $\mathcal{H}om(\mathcal{C},\mathcal{C}')$. We define for $A \in \mathcal{C}$:

$$\ker(f)(A) = \ker(F(A) \xrightarrow{f(A)} G(A)).$$

Together with the natural transformation $\ker(f) \to F$ given by $\ker(f(A)) \to F(A)$, the functor $\ker(f)$ is the kernel of $f : F \to G$ in the category $\mathcal{H}om(\mathcal{C},\mathcal{C}')$.

Similarly, by defining $\mathrm{coker}(f)(A) = \mathrm{coker}(F(A))$, $\mathrm{im}(f)(A) = \mathrm{im}(f(A))$ and $\mathrm{coim}(f)(A) = \mathrm{coim}(f(A))$ we obtain the cokernel, the image and the coimage of f in the category $\mathcal{H}om(\mathcal{C},\mathcal{C}')$ and therefore at once the isomorphy of the canonical morphism $\bar{f} : \mathrm{coim}(f) \to \mathrm{im}(f)$. Hence $\mathcal{H}om(\mathcal{C},\mathcal{C}')$ is an abelian category. Furthermore, the exactness condition on $F' \to F \to F''$ in $\mathcal{H}om(\mathcal{C},\mathcal{C}')$, mentioned in the proposition, is obvious from the description of the kernel and the image just given. □

(1.4) Injective Objects

For each object M in an abelian category $\mathcal{C}$ the contravariant functor $A \mapsto \mathrm{Hom}(A, M)$ from $\mathcal{C}$ to the category of abelian groups is always left exact. M is called an **injective object** if this functor is exact, hence if each morphism $A' \to M$ of a subobject A' of A extends to a morphism $A \to M$.

An abelian category $\mathcal{C}$ is said to have **sufficiently many injective objects** if for each object $A \in \mathcal{C}$ there exists a monomorphism of A into an injective object of $\mathcal{C}$. It is interesting to know under which conditions an abelian category possesses sufficiently many injective objects.

Let us assume the abelian category $\mathcal{C}$ has the property:

Ab 3) For each family $(A_i)_{i \in I}$ of objects A_i in $\mathcal{C}$ the direct sum of the A_i exists.

In particular then, for each family of subobjects A_i of $A \in \mathcal{C}$ the object $\sum_i A_i = \mathrm{im}(\bigoplus_i A_i \to A)$ exists, where the morphism $\bigoplus_i A_i \to A$ is induced by the canonical injections $A_i \to A$.

With this notation we formulate the following axiom:

Ab 5) Assume Ab 3) holds and that for each increasingly filtered family of subobjects A_i of $A \in \mathcal{C}$ and for each system of morphisms $u_i : A_i \to B$ into a fixed object $B \in \mathcal{C}$, such that u_i is induced by u_j if $A_i \subset A_j$, there is a (unique) morphism $u : \sum_i A_i \to B$ inducing the u_i.

Remark. For a thorough discussion of the axioms Ab 3) and Ab 5) and of other axioms (in particular Ab 4)) see [14], ch. I, 1.5.

We also need the following notion: A family $(Z_i)_{i \in I}$ of objects in an arbitrary category $\mathcal{C}$ is called a **family of generators of** $\mathcal{C}$ if for each $A \in \mathcal{C}$ and each subobject B of A, $B \neq A$, there exists an $i \in I$ and a morphism $Z_i \to A$ which does not factor through the canonical inclusion $B \to A$. An object $Z \in \mathcal{C}$ is a **generator of** $\mathcal{C}$ if the family consisting only of Z is a family of generators. We have:

(1.4.1) Proposition. *Let $\mathcal{C}$ be an abelian category satisfying Ab 3), let $(Z_i)_{i \in I}$ be a family of objects in $\mathcal{C}$ and let $Z = \bigoplus_i Z_i$. The following are equivalent:*

a) $(Z_i)_{i \in I}$ is a family of generators of $\mathcal{C}$.

b) Z is a generator of $\mathcal{C}$.

c) For each $A \in \mathcal{C}$ there is an exact sequence of the form $\bigoplus Z \to A \to 0$.

For the simple proof of this proposition see [14], ch. I, 1.9.1.

(1.4.2) Theorem. *Let $\mathcal{C}$ be an abelian category. If $\mathcal{C}$ satisfies Ab 5) and if $\mathcal{C}$ has generators, then $\mathcal{C}$ has sufficiently many injective objects.*

This theorem is proved in [14], ch. I, 1.10.

A first example of an abelian category satisfying Ab 5) and having generators is the category *Ab* of abelian groups. More examples are obtained as follows:

(1.4.3) Proposition. *Let $\mathcal{C}$ and $\mathcal{C}'$ be categories with $\mathcal{C}'$ abelian. Consider the category $\mathcal{H}om(\mathcal{C},\mathcal{C}')$ of functors from $\mathcal{C}$ to $\mathcal{C}'$ (compare (1.3.1)). Then*

a) If $\mathcal{C}'$ satisfies Ab 5), so does $\mathcal{H}om(\mathcal{C},\mathcal{C}')$.

b) If $\mathcal{C}'$ satisfies Ab 3) and has generators, both properties also hold in $\mathcal{H}om(\mathcal{C},\mathcal{C}')$.

Proof: a) Given a family (F_i) of objects in $\mathcal{H}om(\mathcal{C},\mathcal{C}')$ the functor $F : \mathcal{C} \to \mathcal{C}'$ defined by $F(A) = \bigoplus_i F_i(A)$, together with the natural transformations $F_i \to F$ defined by the injections $F_i(A) \to F(A)$, is the direct sum of the F_i in $\mathcal{H}om(\mathcal{C},\mathcal{C}')$.

Now let (F_i) be a family of subobjects of $F \in \mathcal{H}om(\mathcal{C},\mathcal{C}')$ and let $f_i : F_i \to G$ be a family of morphisms in $\mathcal{H}om(\mathcal{C},\mathcal{C}')$, s.t. f_i is induced from f_j in case $F_i \subset F_j$. Assume furthermore that the family (F_i) is increasingly filtered. For the object $\sum_i F_i = \operatorname{im}(\bigoplus_i F_i \to F)$ we get $(\sum_i F_i)(A) = \sum_i F_i(A)$ for each $A \in \mathcal{C}$ (compare the proof of (1.3.1)). Each $F_i(A)$ gets identified with a subobject of $F(A)$ and the family $(F_i(A))$ is increasingly filtered. Furthermore we have morphisms $f_i(A) : F_i(A) \to G(A)$, s.t. $f_i(A)$ is induced from $f_j(A)$ in case $F_i(A) \subset F_j(A)$. Since $\mathcal{C}'$ has property Ab 5) there is then a uniquely determined morphism $f(A) : (\sum_i F_i)(A) = \sum_i f_i(A) \to G(A)$ inducing the $f_i(A)$. Obviously this morphism $f(A)$ is functorial in A, and hence we get a (unique) morphism $f : \sum_i F_i \to G$ inducing the f_i. Therefore $\mathcal{H}om(\mathcal{C},\mathcal{C}')$ satisfies Ab 5).

b) The transition of property Ab 3) from $\mathcal{C}'$ to $\mathcal{H}om(\mathcal{C},\mathcal{C}')$ was already shown at the beginning of part a).

Let Z be a generator of $\mathcal{C}'$ (see (1.4.1)). For each $A \in \mathcal{C}$ we define an object Z_A in $\mathcal{H}om(\mathcal{C},\mathcal{C}')$, i.e. a functor $Z_A : \mathcal{C} \to \mathcal{C}'$, via

$$Z_A(B) = \bigoplus_{\operatorname{Hom}(B,A)} Z$$

for $B \in \mathcal{C}$ with the obvious morphisms $Z_A(u)$ for $B \xrightarrow{u} B'$ in $\mathcal{C}$. We claim that the family $(Z_A)_{A\in\mathcal{C}}$ is a family of generators for the category $\mathcal{H}om(\mathcal{C},\mathcal{C}')$.

First of all we have for each object A in $\mathcal{C}$ and each F in $\mathcal{H}om(\mathcal{C},\mathcal{C}')$ a canonical isomorphism

$$(*) \qquad \operatorname{Hom}(Z, F(A)) \xrightarrow{\sim} \operatorname{Hom}(Z_A, F)$$

by assigning to each morphism $v' : Z \to F(A)$ in $\mathcal{C}'$ the natural transformation $V' : Z_A \to F$, where $V'(B) : Z_A(B) = \bigoplus_{\operatorname{Hom}(A,B)} Z \to F(B)$

is induced from $Z \xrightarrow{v'} F(A) \xrightarrow{F(u)} F(B)$ for $A \xrightarrow{u} B$. The inverse map assigns to the natural transformation $V' : Z_A \to F$ the morphism $Z \to \bigoplus_{\mathrm{Hom}(A,A)} Z = Z_A(A) \xrightarrow{V'(A)} F(A)$, where the first arrow is the injection of Z into $\bigoplus Z$ determined by the identical morphism id_A of A.

Now, given $F \in \mathcal{H}om(\mathcal{C},\mathcal{C}')$ and a subobject G of F, $G \neq F$, the object $G(A)$ is a proper subobject of $F(A)$ for at least one A in $\mathcal{C}$. Therefore there is a morphism $Z \to F(A)$ in $\mathcal{C}'$ that does not factor through the canonical injection $G(A) \to F(A)$. Obviously then, the morphism $Z_A \to F$ corresponding to $Z \to F(A)$ via the isomorphism $(*)$ has the property that it does not factor through the injection $G \to F$. □

§ 2. Homological Algebra in Abelian Categories

(2.1) ∂-Functors

Let $\mathcal{C}$ be an abelian and let $\mathcal{C}'$ be an additive category. A **covariant ∂-functor from $\mathcal{C}$ to $\mathcal{C}'$** is a system $T = (T^i)_{i\geq 0}$ of covariant additive functors

$$T^i : \mathcal{C} \to \mathcal{C}'$$

together with a "connecting" morphism

$$\partial : T^i(A'') \to T^{i+1}(A')$$

defined for each $i \geq 0$ and each short exact sequence $0 \to A' \to A \to A'' \to 0$ in $\mathcal{C}$, satisfying the following properties:

i) Given a commutative diagram with exact rows

$$\begin{array}{ccccccccc} 0 & \longrightarrow & A' & \longrightarrow & A & \longrightarrow & A'' & \longrightarrow & 0 \\ & & \downarrow & & \downarrow & & \downarrow & & \\ 0 & \longrightarrow & B' & \longrightarrow & B & \longrightarrow & B'' & \longrightarrow & 0 \end{array}$$

in $\mathcal{C}$, the diagram

$$\begin{array}{ccc} T^i(A'') & \xrightarrow{\partial} & T^{i+1}(A') \\ \downarrow & & \downarrow \\ T^i(B'') & \xrightarrow{\partial} & T^{i+1}(B') \end{array}$$

is commutative for all $i \geq 0$.

ii) Given an exact sequence $0 \to A' \to A \to A'' \to 0$ in $\mathcal{C}$ the long sequence $0 \to T^0(A') \to T^0(A) \to T^0(A'') \to T^1(A') \to T^1(A) \to \cdots$ is a complex in $\mathcal{C}'$.

If the category $\mathcal{C}'$ is abelian as well, a ∂-functor $T = (T^i)_{i \geq 0}$ from $\mathcal{C}$ to $\mathcal{C}'$ is **exact**, if for each exact sequence $0 \to A' \to A \to A'' \to 0$ in $\mathcal{C}$ the corresponding long sequence

$$\cdots \to T^i(A) \to T^i(A'') \to T^{i+1}(A') \to \cdots$$

is exact.

Let $T = (T^i)_{i \geq 0}$ and $T' = (T'^i)_{i \geq 0}$ be two ∂-functors from the abelian category $\mathcal{C}$ to the additive category $\mathcal{C}'$. A **morphism from T to T'** is a system $f = (f^i)_{i \geq 0}$ of functorial morphisms

$$f^i : T^i \to T'^i$$

which commute naturally with ∂. That is, for any exact sequence $0 \to A' \to A \to A'' \to 0$ in $\mathcal{C}$ the diagram

$$\begin{array}{ccc} T^i(A'') & \longrightarrow & T^{i+1}(A') \\ {\scriptstyle f^i(A'')}\downarrow & & \downarrow{\scriptstyle f^{i+1}(A')} \\ T'^i(A'') & \longrightarrow & T'^{i+1}(A') \end{array}$$

commutes.

A ∂-functor $T = (T^i)_{i \geq 0}$ from the abelian category $\mathcal{C}$ to the additive category $\mathcal{C}'$ is called **universal** if each morphism $f^0 : T^0 \to T'^0$ of functors has one and only one extension to a morphism $f : T \to T'$.

For example, if $F : \mathcal{C} \to \mathcal{C}'$ is a covariant additive functor, there is up to a canonical isomorphism at most one universal ∂-functor T from $\mathcal{C}$ to $\mathcal{C}'$ with $T^0 = F$.

An additive covariant functor $F : \mathcal{C} \to \mathcal{C}'$ from the abelian category $\mathcal{C}$ to the additive category $\mathcal{C}'$ is called **effaceable** if for each object A in $\mathcal{C}$ there is a monomorphism $u : A \to M$ in $\mathcal{C}$ such that $F(u) = 0$.

(2.1.1) Lemma. *If $\mathcal{C}$ is an abelian category with sufficiently many injective objects (compare (1.6)), a functor $F : \mathcal{C} \to \mathcal{C}'$ is effaceable if and only if $F(M) = 0$ for all injective objects M in $\mathcal{C}$.*

Proof: Assume F is effaceable and let M be an injective object in $\mathcal{C}$. By definition, there exists a monomorphism $u : M \to N$ in $\mathcal{C}$ with $F(u) = 0$. Since M is injective there is a morphism $v : N \to M$, s.t. $vu = 1_M$. This implies $1_{F(M)} = F(1_M) = F(v)F(u) = 0$, hence $F(M) = 0$. The converse is even more obvious. □

(2.1.2) Theorem. *Let $\mathcal{C}$ be an abelian category with sufficiently many injective objects and let $T = (T^i)_{i \geq 0}$ be an exact ∂-functor from $\mathcal{C}$ to an abelian category $\mathcal{C}'$. Then the following are equivalent:*

a) T is universal.

b) T^i is effaceable for $i > 0$.

A proof of this theorem can be found in [14], ch. I, 2.2.1. The implication b) $\Longrightarrow$ a) holds without the assumption that $\mathcal{C}$ has sufficiently many injective objects.

(2.2) Derived Functors

Let $F : \mathcal{C} \to \mathcal{C}'$ be a left exact additive covariant functor between abelian categories. The **right derived functor of** F is the universal exact ∂-functor from $\mathcal{C}$ to $\mathcal{C}'$ extending F. If it exists, it is unique up to a canonical isomorphism. It is then denoted by $(R^iF)_{i \geq 0}$, and R^iF is called the i-th right derived functor of F.

(2.2.1) Theorem. *Let $\mathcal{C}$ be an abelian category with sufficiently many injective objects and let $\mathcal{C}'$ be an abelian category. Then for each left exact additive covariant functor $F : \mathcal{C} \to \mathcal{C}'$ the right derived functor $(R^iF)_{i \geq 0}$ exists.*

Remark. For a more general result see [14], ch. I, th. 2.2.2.

Proof of theorem (2.2.1): Since $\mathcal{C}$ has sufficiently many injective objects each object $A \in \mathcal{C}$ has an **injective resolution**, i.e. there is an exact sequence

$$M^*(A): \quad 0 \to A \to M^0 \to M^1 \to \cdots$$

with injective objects M^i in $\mathcal{C}$.

Now we use the following two facts (cp. [5], ch. V):

a) If $M^*(A)$ and $M^*(A')$ are injective resolutions of A and A' in $\mathcal{C}$, then each morphism $u : A \to A'$ extends to a morphism $M^*(A) \to M^*(A')$ of complexes and every two extensions of u are homotopic. In particular, the injective resolution $M^*(A)$ of A is uniquely determined up to homotopy.
b) Any exact sequence $0 \to A' \to A \to A'' \to 0$ in $\mathcal{C}$ is extendable to an exact sequence

$$0 \to M^*(A') \to M^*(A) \to M^*(A'') \to 0$$

of injective resolutions of A', A and A''.

Given an object A in $\mathcal{C}$, apply the functor F to an injective resolution

$$M^*(A): \quad 0 \to A \to M^0 \to M^1 \to \cdots$$

of A and define

$$R^0F(A) = \ker(F(M^0) \to F(M^1))$$
$$R^iF(A) = \frac{\ker(F(M^i) \to F(M^{i+1}))}{\operatorname{im}(F(M^{i-1}) \to F(M^i))}, \quad i \geq 1.$$

By a) $R^iF(A)$ is well-defined (i.e. independent of the injective resolution of A chosen), and moreover we have for each morphism $u : A \to A'$ in $\mathcal{C}$ a morphism $R^iF(u) : R^iF(A) \to R^iF(A')$. The functors $R^iF : \mathcal{C} \to \mathcal{C}'$ are additive. Since F is left exact, $R^0F = F$. For $i > 0$ the R^iF are effaceable, since for an injective $M \in \mathcal{C}$ an injective resolution of M is given by $0 \to M \xrightarrow{\equiv} M \to 0$, from which we read off $R^iF(M) = 0$ for $i > 0$. Furthermore, if $0 \to A' \to A \to A'' \to 0$ is a short exact sequence in $\mathcal{C}$, it is extendable by b) to a short exact sequence

$$0 \to M^*(A') \to M^*(A) \to M^*(A'') \to 0$$

for suitably chosen resolutions $M^*(A'), \ldots$. Since $M^i(A')$ is injective, all exact sequences

$$0 \to M^i(A') \to M^i(A) \to M^i(A'') \to 0$$

split and therefore

$$0 \to F(M^i(A')) \to F(M^i(A)) \to F(M^i(A'')) \to 0$$

remains exact. The exact sequence

$$0 \to F(M^*(A')) \to F(M^*(A)) \to F(M^*(A'')) \to 0$$

of complexes in $\mathcal{C}'$ yields, as usual, the connecting homomorphisms

$$\partial : R^i F(A'') \to R^{i+1} F(A')$$

of the cohomology groups, so that the long cohomology sequence becomes exact and the ∂'s functorial for short exact sequences in $\mathcal{C}$. □

(2.3) Spectral Sequences

Let A be an object of the abelian category $\mathcal{C}$. A (decreasing) **filtration of** A is a family $(F^p(A))_{p\in\mathbb{Z}}$ of subobjects $F^p(A)$ of A, such that $F^{p+1}(A) \subset F^p(A)$ for all p. One defines $gr_p(A) = F^p(A)/F^{p+1}(A)$. Given filtered objects A and B in $\mathcal{C}$ a morphism $u : A \to B$ is said to be compatible with the filtration if $u(F^p(A)) \subset F^p(B)$ for all $p \in \mathbb{Z}$.

A **spectral sequence** in the abelian category $\mathcal{C}$ is a system

$$E = (E_r^{pq}, E^n)$$

consisting of:

a) objects $E_r^{pq} \in \mathcal{C}$ for all $(p,q) \in \mathbb{Z} \times \mathbb{Z}$ and $r \geq 2$,
b) morphisms $d_r^{pq} : E_r^{pq} \to E_r^{p+r,q-r+1}$ in $\mathcal{C}$ such that $d_r^{p+r,q-r+1} \circ d_r^{pq} = 0$,
c) isomorphisms $\alpha_r^{pq} : \ker(d_r^{pq})/\mathrm{im}(d_r^{p-r,q-r+1}) \xrightarrow{\cong} E_{r+1}^{pq}$,
d) decreasingly filtered objects $E^n \in \mathcal{C}$ for all $n \in \mathbb{Z}$. It will be assumed that for each fixed pair $(p,q) \in \mathbb{Z} \times \mathbb{Z}$ the morphisms d_r^{pq} and $d_r^{p-r,q+r-1}$ vanish for sufficiently large r. It follows from c) that the objects E_r^{pq} are then independent of r for r sufficiently large, and we denote this object by E_∞^{pq}. It will also be assumed that for each $n \in \mathbb{Z}$ the following holds: $F^p(E^n) = E^n$ for sufficiently small p and $F^p(E^n) = 0$ for sufficiently large p.
e) isomorphisms $\beta^{pq} : E_\infty^{pq} \xrightarrow{\cong} gr_p(E^{p+q})$.

Remarks. For a spectral sequence $E = (E_r^{pq}, E^n)$ in the abelian category $\mathcal{C}$, the E_2^{pq} are called the **initial terms** and the E^n the **limit terms** of the spectral sequence. Most of the time we write $E_2^{pq} \Longrightarrow E^{p+q}$ for $E = (E_r^{pq}, E^n)$. – For a more general definition of spectral sequences compare [30], 0_{III}, 11.1.

A morphism

$$u : E = (E_r^{pq}, E^n) \to E' = (E'^{pq}_r, E'^n)$$

of spectral sequences in $\mathcal{C}$ is a system of morphisms

$$\begin{cases} u_r^{pq}: & E_r^{pq} \to E'^{pq}_r \\ u^n: & E^n \to E'^n, \end{cases}$$

where the u^n are compatible with the filtrations of E^n und E'^n and the u_r^{pq} resp. u^n naturally commute with d_r^{pq}, α_r^{pq} and β^{pq}.

The spectral sequences E in $\mathcal{C}$ form an additive category. An additive functor from an abelian category to a category of spectral sequences is called a **spectral functor**.

A spectral sequence $E = (E_r^{pq}, E^n)$ in $\mathcal{C}$ is called a **cohomological spectral sequence** if $E_2^{pq} = 0$ for $p < 0$ and $q < 0$.

(2.3.1) Proposition. *For each cohomological spectral sequence $E = (E_r^{pq}, E^n)$ in $\mathcal{C}$ there are morphisms*

$$\begin{cases} E_2^{n,0} & \to E^n \\ E^n & \to E_2^{0,n} \end{cases}$$

that are functorial in E, the so-called **edge morphisms**.

Proof: Since $E_2^{pq} = 0$ for $p < 0$ or $q < 0$ the defining property c) for spectral sequences implies that for $p < 0$ or $q < 0$ we have $E_r^{pq} = 0$ for $r \geq 2$ and hence also $E_\infty^{pq} = 0$.

This has the following consequences for the filtration $F^p(E^n)$ of the limit terms E^n:

$$\begin{cases} F^0(E^n) & = E^n \\ F^{n+1}(E^n) & = 0 \end{cases}$$

In fact, for each $i > 0$ we have $gr_{-i}(E^n) \cong E_\infty^{-i,n-i} = 0$, hence $gr_{-1}(E^n) = F^{-1}(E^n)/F^0(E_n) = 0$, $gr_{-2}(E^n) = F^{-2}(E^n)/F^{-1}(E^n) = 0$, i.e. $F^0(E^n) = F^{-1}(E^n) = F^{-2}(E^n) = \cdots = E^n$. Furthermore, for each $i > 0$ we have $gr_{n+i}(E^n) \cong E_\infty^{n+i,-i} = 0$, hence $gr_{n+1}(E^n) = F^{n+1}(E^n)/(F^{n+2}(E^n)) = 0, \ldots$, i.e. $F^{n+1}(E^n) = F^{n+2}(E^n) = \ldots = 0$. The edge morphisms are now constructed in the following way: Since F^{n+1} $(E^n) = 0$, we have the monomorphism

(i) $$E_\infty^{n,0} \cong F^n(E^n)/F^{n+1}(E^n) \cong F^n(E^n) \to E^n.$$

For $r \geq 2$ we have

$$E_{r+1}^{n,0} \cong \frac{\ker(E_r^{n,0} \to E_r^{n+r,-r+1})}{\operatorname{im}(E_r^{n-r,r-1} \to E_r^{n,0})}.$$

Since $E_r^{n+r,-r+1} = 0$, we obtain an epimorphism $E_r^{n,0} \to E_{r+1}^{n,0}$, hence an epimorphism

(ii) $$E_2^{n,0} \to E_\infty^{n,0}.$$

The composite of the morphisms in (ii) and (i) is the first edge morphism $E_2^{n,0} \to E^n$. Moreover, since $F^0(E^n) = E^n$ we have an epimorphism

(i') $$E^n = F^0(E^n) \to F^0(E^n)/F'(E^n) \overset{\cong}{\to} E_\infty^{0,n}.$$

For $r \geq 2$:

$$E_{r+1}^{0,n} \cong \frac{\ker(E_r^{0,n} \to E_r^{r,n-r+1})}{\operatorname{im}(E_r^{-r,n+r-1} \to E_r^{0,r})}.$$

Now $E_r^{-r,n+r-1} = 0$, and we obtain a monomorphism $E_{r+1}^{0,n} \to E_r^{0,n}$, hence a monomorphism

(ii') $$E_\infty^{0,n} \to E_2^{0,n}.$$

The composite of the morphisms in (i') and (ii') is the second edge morphism $E^n \to E_2^{0,n}$. The functoriality of the edge morphism is immediate from the construction. □

(2.3.2) Proposition. *For a cohomological spectral sequence $E_2^{pq} \Longrightarrow E^{p+q}$ the sequence*

$$0 \to E_2^{1,0} \to E^1 \to E_2^{0,1} \overset{d_2}{\to} E_2^{2,0} \to E^2$$

is exact. It is called **the exact sequence of terms of low degree** *(or* **five term exact sequence***) belonging to $E_2^{pq} \Longrightarrow E^{p+q}$.*

For a proof see [5], ch. XV, § 5.

Under additional assumptions on the spectral sequence, it is possible to obtain results on the edge morphisms between terms of higher degree. We mention the following result (cp. [5], ch. XV, § 5):

(2.3.3) Proposition. *Assume that for a cohomological spectral sequence $E_2^{pq} \Longrightarrow E^{p+q}$ the terms E_2^{pq} vanish for $0 < q < n$. Then*

$$E_2^{m,0} \cong E^m \quad \textit{for } m < n,$$

and the sequence

$$0 \to E_2^{n,0} \to E^n \to E_2^{0,n} \to E_2^{n+1,0} \to E_2^{n+1}$$

is exact.

(2.3.4) Corollary. *If $E_2^{pq} = 0$ for all $q > 0$, we have*

$$E_2^{n,0} \cong E^n$$

for all n. The spectral sequence $E_2^{p,q} \Longrightarrow E^{p+q}$ is then called **trivial**.

Remark. Under similar conditions on p, symmetric to those on q, in the spectral sequence $E_2^{pq} \Longrightarrow E^{p+q}$, analogous results can be obtained on the edge morphisms $E^n \to E_2^{0,n}$.

One of the most important examples of spectral sequences is the spectral sequence for the right derived functor of a composite of functors:

(2.3.5) Theorem. *Let $\mathcal{C}$ and $\mathcal{C}'$ be abelian categories with sufficiently many injective objects and let $\mathcal{C}''$ be another abelian category. Let $F : \mathcal{C} \to \mathcal{C}'$ and $G : \mathcal{C}' \to \mathcal{C}''$ be left exact additive covariant functors. Assume the functor F maps injective objects from $\mathcal{C}$ to* **G-acyclic** *objects (i.e. those annihilated by R^qG for $q > 0$) in $\mathcal{C}'$. Then there is a cohomological spectral functor*

$$A \mapsto (E_2^{pq}(A) \Longrightarrow E^{p+q}(A))$$

from $\mathcal{C}$ to the category of spectral sequences in $\mathcal{C}''$, such that

$$\begin{cases} E_2^{pq}(A) & = R^pG(R^qF(A)) \\ E^n(A) & = R^n(G \circ F)(A) \end{cases}$$

For the proof of this theorem compare [14], ch. I, 2.4.1.

Under the assumptions of the theorem there are, for each object $A \in \mathcal{C}$, the edge morphisms

$$\begin{cases} R^nG(F(A)) & \to R^n(G \circ F)(A) \\ R^n(G \circ F)(A) & \to G(R^nF(A)), \end{cases}$$

which are functorial in A. Here, the edge morphisms $R^n(G \circ F)(A) \to G(R^nF(A))$ have the following interpretation: The system $(G \circ R^nF)_{n \geq 0}$ of covariant additive functors $G \circ R^n F : \mathcal{C} \to \mathcal{C}''$ is naturally a ∂-functor from $\mathcal{C}$ to $\mathcal{C}''$. Then the universality of the right derived functor $(R^n(G \circ F))_{n \geq 0}$ of $G \circ F : \mathcal{C} \to \mathcal{C}''$ implies that there is exactly one morphism of ∂-functors

$$(R^n(G \circ F))_{n \geq 0} \to (G \circ R^n F)_{n \geq 0}$$

extending the identity morphism $G \circ F \to G \circ F$, and this is the second edge morphism.

§ 3. Inductive Limits

(3.1) Limit Functors

Let $\mathcal{I}$ and $\mathcal{C}$ be categories. Attached to each object X in $\mathcal{C}$ we have the constant functor $c_X : \mathcal{I} \to \mathcal{C}$ with value X, which maps each object from $\mathcal{I}$ to X and each morphism in $\mathcal{I}$ to the arrow id_X. Moreover, any morphism $X \to Y$ in $\mathcal{C}$ induces in an obvious way a morphism $c_X \to c_Y$ of functors.

Let $F : \mathcal{I} \to \mathcal{C}$ be a functor. We obtain a covariant functor

$$\underrightarrow{lim}\, F : \mathcal{C} \to sets$$

by assigning to each $X \in \mathcal{C}$ the set $\mathrm{Hom}(F, c_X)$ of natural transformations $F \to c_X$. If $\underrightarrow{lim}\, F$ is a representable functor, the representing object in $\mathcal{C}$ is denoted by $\varinjlim F$ (more precisely: $\varinjlim_{\mathcal{I}} F$ or $\varinjlim_{i \in \mathcal{I}} F(i)$) and called the **inductive limit of** F.

Dually, we have a contravariant functor

$$\underleftarrow{lim}\, F : \mathcal{C} \to sets$$

via $X \to \mathrm{Hom}(c_X, F)$. If $\underleftarrow{lim}\, F$ is representable, the representing object in $\mathcal{C}$ is denoted by $\varprojlim F$ and called the **projective limit of** F.

Given functors $F, F' : \mathcal{I} \to \mathcal{C}$, each natural transformation $f : F \to F'$ induces a natural transformation $\underrightarrow{lim}\, f : \underrightarrow{lim}\, F \to \underrightarrow{lim}\, F'$. That is, $\underrightarrow{lim}$ is a contravariant functor from the category $\mathcal{H}om(\mathcal{I}, \mathcal{C})$ to the category $\mathcal{H}om(\mathcal{C}, sets)$. If $\underrightarrow{lim}\, F$ and $\underrightarrow{lim}\, F'$ are representable, there is a unique morphism $\varinjlim f : \varinjlim F \to \varinjlim F'$ in $\mathcal{C}$ corresponding to the natural transformation $\varinjlim f$.

(3.1.1) Proposition. *Let $\mathcal{C}$ be an abelian category satisfying Ab 3) (i.e. having arbitrary direct sums). Then $\varinjlim F$ is representable for each functor $F : \mathcal{I} \to \mathcal{C}$, and*

$$\varinjlim : \mathcal{H}om(\mathcal{I}, \mathcal{C}) \to \mathcal{C}$$

is a right exact additive functor of abelian categories (cp. (1.3.1)).

Proof: The construction of the object in $\mathcal{C}$ representing $\varinjlim F$ is done in a standard way: We form the direct sum $S = \bigoplus F(i)$ in $\mathcal{C}$, where the summation extends over all objects $i \in \mathcal{I}$. For each morphism $u : i \to j$, the subobject $N_{i \to j}$ of S is defined as the image of the morphism $F(i) \to S$, which is induced by the morphism $F(i) \to F(i) \oplus F(j)$ with components $\mathrm{id}_{F(i)}$ and $F(-u)$. Next we form the subobject $N = \sum N_{i \to j} = \mathrm{im}(\bigoplus N_{i \to j} \to S)$ of S where the summation extends over all morphisms $i \to j$ of $\mathcal{I}$. The quotient S/N together with the obvious morphisms $F(i) \to S/N$ has the desired properties. The other statements of the proposition are easily checked. □

(3.1.2) Remark. If we choose for $\mathcal{C}$ the category *sets* of sets, $\varinjlim F$ is representable for each functor $F : \mathcal{I} \to$ *sets* in this case as well: We form the disjoint union S of the sets $F(i)$ for $i \in I$, and divide by the equivalence relation N generated by all pairs (x, y) of elements $x \in F(i)$ and $y \in F(j)$ for which there is a morphism $u : i \to j$ in $\mathcal{I}$, such that $F(u)(x) = y$. S/N together with $F(i) \to S/N$ has the desired properties. For a statement generalizing both (3.1.1) and (3.1.2) compare [2], exp. I, 2.2.1 and 2.3.1.

(3.1.3) Proposition. *Let $F : \mathcal{C} \to \mathcal{C}'$ be a functor between arbitrary categories, which has a left adjoint functor ${}^{ad}F : \mathcal{C}' \to \mathcal{C}$ (cp. (1.1)). Then F commutes with representable projective limits and ${}^{ad}F$ commutes with representable inductive limits.*

For ${}^{ad}F$ (and similarly for F), this is supposed to mean the following: If $G : \mathcal{I} \to \mathcal{C}'$ is a functor with representable inductive limit, the same is true for the composite ${}^{ad}F \circ G : \mathcal{I} \to \mathcal{C}$, and we have a canonical isomorphism

$$\varinjlim({}^{ad}F \circ G) \xrightarrow{\cong} {}^{ad}F(\varinjlim G)$$

(cp. [2], exp. I, 2.11.)

(3.2) Exactness of Inductive Limits

A category $\mathcal{I}$ is called **pseudofiltered** if the two following properties hold:

PS 1) Each diagram of the form

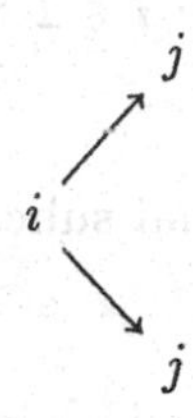

in $\mathcal{I}$ can be extended to a commutative diagram of the form

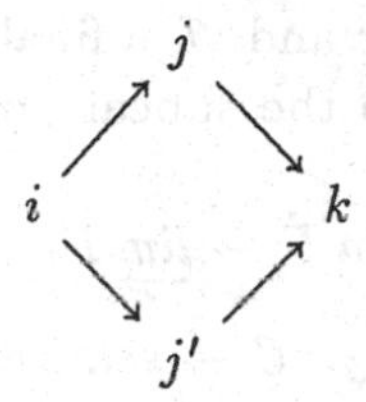

PS 2) Each diagram of the form

$$i \overset{u}{\underset{v}{\rightrightarrows}} j$$

can be extended to a diagram

$$i \overset{u}{\underset{v}{\rightrightarrows}} j \xrightarrow{w} k$$

such that $wu = wv$.

A non-empty category $\mathcal{I}$ is called **filtered** if it is pseudofiltered, and for each two objects i and j there is an object k together with morphisms $i \to k$ and $j \to k$. Example: If amalgamated sums (resp. sums of two objects) and cokernels of double arrows exist in $\mathcal{I}$, then $\mathcal{I}$ is pseudofiltered (resp. filtered).

(3.2.1) Theorem. *Let $\mathcal{I}$ be a pseudofiltered category and $\mathcal{C}$ an abelian category satisfying Ab 5). Then the functor*

$$\varinjlim : \mathcal{H}om(\mathcal{I},\mathcal{C}) \to \mathcal{C}$$

is exact.

For a proof see [30], 14.6.6.

(3.3) Final Subcategories.

Let $\mathcal{I}$ be a category. A subcategory $\mathcal{J}$ of $\mathcal{I}$ is called **final** if $\mathcal{J}$ is a full subcategory of $\mathcal{I}$ and for every $i \in \mathcal{I}$ there is a morphism $i \to j$ in $\mathcal{I}$ with $j \in \mathcal{J}$.

If $\mathcal{I}$ is pseudofiltered, every final subcategory of $\mathcal{I}$ is obviously pseudofiltered as well.

If $\mathcal{I}$ has a final object ∞, $\{\infty, \mathrm{id}_\infty\}$ is a final subcategory of $\mathcal{I}$.

Let $F : \mathcal{I} \to \mathcal{C}$ be a functor and $\mathcal{J}$ a final subcategory of $\mathcal{I}$. Let $F|_\mathcal{J}$ denote the restriction of F to the subcategory $\mathcal{J}$. There is a canonical morphism

$$\varinjlim F \to \varinjlim F|_\mathcal{J}$$

for the functors $\varinjlim F$, $\varinjlim F|_\mathcal{J} : \mathcal{C} \to sets$, and the following holds:

(3.3.1) Proposition. *If the category $\mathcal{I}$ satisfies PS 1) (cp. (3.2)), the morphism*

$$\varinjlim F \to \varinjlim F|_\mathcal{J}$$

is an isomorphism for every final subcategory $\mathcal{J}$ of $\mathcal{I}$.

Proof: We have to show that for each $X \in \mathcal{C}$ the map $\mathrm{Hom}(F, c_X) \to \mathrm{Hom}(F|_\mathcal{J}, c_X)$ is bijective. This map is injective: Assume that $\bar{\varphi}, \bar{\psi} : F \to c_X$ are natural transformations s.t. $\bar{\varphi}(j) = \bar{\psi}(j)$ for all $j \in \mathcal{J}$. Given $i \in \mathcal{I}$, we can find a morphism $i \xrightarrow{u} j$ with $j \in \mathcal{J}$ and therefore $\bar{\varphi}(i) = \bar{\varphi}(j)F(u) = \bar{\psi}(j)F(u) = \bar{\psi}(i)$. The map is also surjective: Given a natural transformation $\varphi : F|_\mathcal{J} \to c_X$ we can define an extension $\bar{\varphi} : F \to c_X$ via $\bar{\varphi}(i) = \varphi(j)F(u)$ for $i \in \mathcal{I}$, where $u : i \to j$ is a morphism from i to an object j in $\mathcal{J}$. Since $\mathcal{I}$ satisfies PS 1), φ is well-defined.

□

In particular, proposition (3.3.1) implies the following:

(3.3.2) If $\mathcal{I}$ has a final object ∞, $\varinjlim F$ is representable for every functor $F : \mathcal{I} \to \mathcal{C}$ and $\varinjlim F \cong F(\infty)$.

Chapter I
Topologies and Sheaves

§ 1. Topologies

(1.1) Preliminaries

Let X be a topological space, and let T denote the family of all open subsets of X. T becomes a category if we define

$$\mathrm{Hom}(U,V) = \begin{cases} \emptyset & \text{if } U \not\subset V \\ \text{inclusion } U \to V & \text{if } U \subset V \end{cases}$$

for $U, V \in T$. X is a final object in the category T. The intersection $\bigcap U_i$ of finitely many $U_1, \ldots, U_n$ in T is equal to the product of the $U_1, \ldots, U_n$ in the category T. The union $\bigcup U_i$ of arbitrarily many U_i in T is equal to the direct sum of the U_i in the category T (cp. 0.1.1).

Let C be either the category of abelian groups or the category of sets. More generally, the following considerations can be based on an arbitrary category with products. A **presheaf** on X with values in C consists of a collection of objects $F(U) \in C$ for $U \in T$ and of a collection of restriction morphisms $res_U^V : F(V) \to F(U)$ for $U \subset V$, such that $res_U^V \circ res_V^W = res_U^W$ for $U \subset V \subset W$ and $res_U^U = id$. In other words, a presheaf on X with values in C is a contravariant functor $F : T \to C$. A morphism $f : F \to G$ of presheaves on X with values in C consists of a collection of morphisms $f(U) : F(U) \to G(U)$ for $U \in T$, such that for each pair U, V in T with $U \subset V$ the diagram

$$\begin{array}{ccc} F(V) & \xrightarrow{f(V)} & G(V) \\ {\scriptstyle res_U^V}\downarrow & & \downarrow{\scriptstyle res_U^V} \\ F(U) & \xrightarrow{f(U)} & G(U) \end{array}$$

commutes. In other words, a morphism of presheaves on X is a morphism $f : F \to G$ of contravariant functors (cp. 0.1.1). A presheaf F on X with values in C is called a **sheaf**, if for every open covering $U = \bigcup U_i$ the diagram

$$F(U) \to \prod_i F(U_i) \rightrightarrows \prod_{i,j} F(U_i \cap U_j),$$

induced by the restriction morphisms, is exact. This means that the first arrow, given by $s \mapsto (res^U_{U_i}(s))$, is injective, and that its image is equal to the kernel of the double arrow, hence consists of all (s_i), s.t. $res^{U_i}_{U_i \cap U_j}(s_i) = res^{U_j}_{U_i \cap U_j}(s_j)$. Morphisms of sheaves are morphisms of presheaves.

Grothendieck's generalisation of the notion of topology – precisely formulated in the next section – consists of replacing the category T of open sets of a topological space X by an arbitrary category, in which e.g. $\mathrm{Hom}(U, V)$ could have more than one element, and of prescribing in addition for this category a system $\{U_i \overset{\varphi_i}{\to} U\}$ of "coverings" of its objects.

(1.2) Grothendieck's Notion of Topology

(1.2.1) Definition. *A* **topology** *(or* **site***) T consists of a category $cat(T)$ and of a set $cov(T)$ of* **coverings**, *i.e. families $\{U_i \overset{\varphi_i}{\to} U\}_{i \in I}$ of morphisms in $cat(T)$, such that the following properties hold:*

(T1) For $\{U_i \to U\}$ in $cov(T)$ and a morphism $V \to U$ in $cat(T)$ all fibre products $U_i \times_U V$ exist and $\{U_i \times_U V \to V\}$ is again in $cov(T)$.

(T2) Given $\{U_i \to U\} \in cov(T)$ and a family $\{V_{ij} \to U_i\} \in cov(T)$ for all $i \in I$, the family $\{V_{ij} \to U\}$, obtained by composition of morphisms, also belongs to $cov(T)$.

(T3) If $\varphi : U' \to U$ is an isomorphism in $cat(T)$ then $\{U' \overset{\varphi}{\to} U\} \in cov(T)$.

(1.2.2) Definition. *A* **morphism** $f : T \to T'$ **of topologies** *is a functor $f : cat(T) \to cat(T')$ of the underlying categories with the following two properties:*

(i) $\{U_i \overset{\varphi_i}{\to} U\} \in cov(T)$ implies $\{f(U_i) \overset{f(\varphi_i)}{\to} f(U)\} \in cov(T')$.

(ii) For $\{U_i \to U\} \in cov(T)$ and a morphism $V \to U$ in $cat(T)$ the canonical morphism

$$f(U_i \times_U V) \to f(U_i) \times_{f(U)} f(V)$$

is an isomorphism for all i.

(1.2.3) Definition. *Let T be a topology and let C denote the category of abelian groups or the category of sets (more generally: a category with products).*

A **presheaf on** T **with values in** $\mathcal{C}$ *is a contravariant functor* $F : T \to \mathcal{C}$.

A **morphism** $f : F \to G$ **of presheaves** *with values in* $\mathcal{C}$ *is defined as a morphism of contravariant functors.*

A presheaf F *is a* **sheaf** *on* T *if for every covering* $\{U_i \to U\}$ *in* T *the diagram*

$$F(U) \to \prod_i F(U_i) \rightrightarrows \prod_{i,j} F(U_i \times_U U_j)$$

is exact in $\mathcal{C}$. *Morphisms of sheaves are defined as morphisms of presheaves.*

Presheaves or sheaves with values in the category of abelian groups are called **abelian presheaves** or **abelian sheaves** on T. From now on we will denote the category of abelian presheaves on T by $\mathcal{P}$ and the category of abelian sheaves on T by $\mathcal{S}$. By definition, $\mathcal{S}$ is a full subcategory of $\mathcal{P}$.

Presheaves or sheaves with values in the category of sets are called presheaves or sheaves of sets for short. We obtain examples of presheaves of sets in the following way: Let Z be an object of the category underlying T. Then $U \mapsto \mathrm{Hom}(U, Z)$ is a presheaf of sets on T. Such presheaves are called **representable presheaves.** It is interesting to know under which conditions representable presheaves are sheaves on T.

(1.3) Examples.

Let X be a topological space. The category of open sets of X together with the usual coverings, i.e. the families $\{U_i \to U\}$ with $U = \bigcup U_i$, defines a topology T.

If $f : X' \to X$ is a continuous map of topological spaces, we obtain a morphism $f^{-1} : T \to T'$ of topologies via $U \mapsto f^{-1}(U)$.

(1.3.1) The canonical topology on a category $\mathcal{C}$ with fibre products

Recall, that a morphism $U \to V$ in $\mathcal{C}$ is called an epimorphism, if the map $\mathrm{Hom}(V, Z) \to \mathrm{Hom}(U, Z)$ is injective for each $Z \in \mathcal{C}$ (cp. 0.1.1). The morphism $U \to V$ is an **effective** epimorphism, if the diagram

$$\mathrm{Hom}(V, Z) \to \mathrm{Hom}(U, Z) \rightrightarrows \mathrm{Hom}(U \times_V U, Z)$$

is exact for each $Z \in \mathcal{C}$. Here the two right-hand maps are induced from the projections of $U \times_V U$ onto the first and second factor. $U \to V$ is called a **universal effective** epimorphism, if $U \times_V V' \to V'$ is an effective epimorphism for each morphism $V' \to V$ in $\mathcal{C}$.

These notions generalize to families of morphisms $U_i \to V$ into a fixed object V: A family $\{U_i \to V\}$ is a family of epimorphisms if

$$\mathrm{Hom}(V, Z) \to \prod_i \mathrm{Hom}(U_i, Z)$$

is injective for each $Z \in \mathcal{C}$. It is a family of effective epimorphisms if the diagram

$$\mathrm{Hom}(V, Z) \to \prod_i \mathrm{Hom}(U_i, Z) \rightrightarrows \prod_{i,j} \mathrm{Hom}(U_i \times_V U_j, Z)$$

is exact for each $Z \in \mathcal{C}$, and finally it is a family of universal effective epimorphisms if $\{U_i \times_V V' \to V'\}$ is a family of effective epimorphisms for each morphism $V' \to V$ in $\mathcal{C}$.

For a more complete treatment of these notions we refer to [7], exp. IV, 1.

On the category $\mathcal{C}$ the **canonical topology** T is now defined by taking for the set of coverings the collection of all families $\{U_i \to U\}$ of universal effective epimorphisms in $\mathcal{C}$. We have to check that the axioms T1) – T3) are satisfied: For T1) and T3) this is trivially the case, and for T2) we only have to extend the result in [7], exp. IV, 1.8, that the composite of universal effective epimorphisms is a universal effective epimorphism, to families of universal effective epimorphisms.

It is immediate from the definition of T that each representable presheaf of sets, i.e. each presheaf of the form $U \mapsto \mathrm{Hom}(U, Z)$ for a fixed Z in $\mathcal{C}$, is a sheaf on T.

Let T' be any other topology on $\mathcal{C}$, such that each representable presheaf of sets on T' is a sheaf on T'. Then every covering $\{U_i \to U\}$ in T' is a family of universal effective epimorphisms in $\mathcal{C}$, and therefore the identity functor on $\mathcal{C}$ is a morphism $T' \to T$ of topologies. Hence the canonical topology is the finest topology on $\mathcal{C}$, in which all representable presheaves of sets are sheaves.

(1.3.2) The canonical topology on the category of left G-sets

Let G be an arbitrary group and let T_G denote the canonical topology on the category of left G-sets with G-maps as morphisms.

We easily check that a family $\{U_i \xrightarrow{\varphi_i} U\}$ of morphisms in the category of left G-sets is a family of universal effective epimorphisms if and only if $U = \bigcup_i \varphi_i(U_i)$.

We know (cp. 1.3.1) that each left G-set Z defines a sheaf of sets on the topology T_G via $U \mapsto \mathrm{Hom}_G(U, Z)$. We want to show that we obtain all sheaves of sets on T_G in this way. More precisely:

(1.3.2.1) Proposition. *The functor $Z \to \mathrm{Hom}_G(\,\cdot\,, Z)$ is an equivalence between the category of left G-sets and the category of sheaves of sets on T_G. The functor $F \to F(G)$ from the category of sheaves of sets on T_G to the category of left G-sets is quasi-inverse to $Z \to \mathrm{Hom}_G(\,\cdot\,, Z)$.*

Here the structure of $F(G)$ as a left G-set is defined as follows: For $g \in G$ and $s \in F(G)$ let $gs = F(\cdot g)(s)$, where $\cdot g : G \to G$ is the G-map $g' \to g'g$.

The proof of 1.3.2.1 runs as follows: The composite of the functors $Z \to \mathrm{Hom}_G(\,\cdot\,, Z)$ and $F \to F(G)$ assigns to each left G-set Z the left G-set $\mathrm{Hom}_G(G, Z)$, which can be canonically identified with Z. The composite of $F \to F(G)$ and $Z \to \mathrm{Hom}_G(\,\cdot\,, Z)$ assigns to each sheaf F the sheaf $\mathrm{Hom}_G(\,\cdot\,, F(G))$. We have to show that there is an isomorphism $F \xrightarrow{\cong} \mathrm{Hom}_G(\,\cdot\,, F(G))$ which is functorial in F. Let U be a left G-set. Then $\{G \xrightarrow{\varphi_u}\}_{u \in U}$ is a covering in the topology T_G, where $\varphi_u(g)$ is defined for each $u \in U$ by $\varphi_u(g) = gu$. For the sheaf F we have the exact diagram

$$F(U) \to \prod_{u \in U} F(G) \rightrightarrows \prod_{u,v \in G} F(G \times_U G)$$

corresponding to the covering. This shows that the image of the injective map $F(U) \to \prod_{u \in U} F(G) = \mathrm{Hom}(U, F(G))$ is precisely the subset $\mathrm{Hom}_G(U, F(G))$ of G-maps $U \to F(G)$. This map

$$F(U) \xrightarrow{\cong} \mathrm{Hom}_G(U, F(G))$$

is functorial in U, hence an isomorphism of sheaves, and it is functorial in F. □

In the special case that A is a left G-module, A defines a sheaf of abelian groups on T_G via $U \mapsto \mathrm{Hom}_G(U, A)$, and we obtain from 1.3.2.1:

(1.3.2.2) Corollary. *The category of left G-modules is equivalent to the category of abelian groups on the canonical topology T_G. The equivalence is given by the mutually quasi-inverse functors $A \to \mathrm{Hom}_G(\,\cdot\,, A)$ and $F \to F(G)$.*

We also note: A homomorphism $G \to H$ of groups induces naturally on each left H-set the structure of a left G-set, and this operation is a morphism $T_H \to T_G$ of topologies (1.2.3).

(1.3.3) The canonical topology on the category of continuous G-sets for a profinite group G

A profinite group G is a topological group, which is the projective limit of finite groups (with the discrete topology) (cp. [32], ch. I, § 1). In a profinite group G the open normal subgroups H form a fundamental system of neighbourhoods of 1, and G gets canonically identified with $\varprojlim G/H$.

A set U, on which G acts continuously from the left (U having discrete topology), is called a **continuous G-set**. The continuity of the G-action is equivalent to the fact that the stabilizer of every element from U is open in G, or that $U = \bigcup U^H$, where H runs through all open normal subgroups of G, and U^H denotes the subset of U of all elements invariant under H.

We consider the category of continuous G-sets with G-maps as morphisms, and we impose on it the canonical topology – again denoted by T_G.

Again we easily check that a family $\{U_i \xrightarrow{\varphi_i} U\}$ of morphisms in the category of continuous G-sets is a family of universal effective epimorphisms if and only if $U = \bigcup_i \varphi_i(U_i)$.

(1.3.3.1) Proposition. *The functor $Z \to \mathrm{Hom}_G(\,\cdot\,, Z)$ is an equivalence between the category of continuous G-sets and the category of sheaves of* *sets on the topology T_G. The quasi-inverse of this functor is the functor* $F \to \varinjlim F(G/H)$.

Here for each open normal subgroup H of G the factor group G/H is viewed as a continuous G-module via left multiplication. We define on the set $F(G/H)$ a continuous G-structure as in the previous example 1.3.2. Moreover, if H and H' are open normal subgroups of G with $H \subset H'$, the canonical G-homomorphism $G/H \to G/H'$ induces a map $F(G/H) \to F(G/H')$. Now the inductive limit considered above is formed over the family of all open normal subgroups H of G, ordered by inclusion. This gives $\varprojlim F(G/H)$ naturally the structure of a continuous G-set.

Proposition (1.3.3.1) is proved as follows: The composite of the functors $Z \to \mathrm{Hom}_G(\,\cdot\,, Z)$ and $F \to \varinjlim F(G/H)$ assigns to each continuous G-set Z the continuous G-set $\varinjlim \mathrm{Hom}_G(G/H, Z)$. For the latter we have canonical identifications: $\varinjlim \mathrm{Hom}_G(G/H, Z) = \varinjlim Z^H = \bigcup Z^H = Z$.

The composite of $F \to \varinjlim F(G/H)$ and $Z \to \mathrm{Hom}_G(\,\cdot\,, Z)$ assigns to each sheaf F of sets on T_G the sheaf $\mathrm{Hom}_G(\,\cdot\,, \varinjlim F(G/H))$, and we obtain an isomorphism of sheaves $F \to \mathrm{Hom}_G(\,\cdot\,, \varinjlim F(G/H))$, functorial in F, in the following way:

Let U be a continuous G-set. Since $U = \bigcup U^H$, the family $\{U^H \to U\}$ of all inclusions $U^H \hookrightarrow U$ is a covering in the topology T_G. From the corresponding exact diagram

$$F(U) \to \prod_H F(U^H) \rightrightarrows \prod_{H,H'} F(U^H \times_U U^{H'})$$

we obtain first of all a canonical identification $F(U) = \varprojlim F(U^H)$, noting that $U^H \times_U U^{H'} = U^H \cap U^{H'}$.

As in the previous example 1.3.2, the exact diagram belonging to the covering $\{G/H \overset{\varphi_u}{\to} U^H\}_{u \in U^H}$, where $\varphi_u(gH) = gu$, induces a canonical isomorphism

$$F(U^H) \overset{\cong}{\to} \mathrm{Hom}_{G/H}(U^H, F(G/H)).$$

We want to show next that the map $F(G/H) \to \varinjlim F(G/H')$ induces a canonical isomorphism

$$\mathrm{Hom}_{G/H}(U^H, F(G/H)) \overset{\cong}{\to} \mathrm{Hom}_G(U^H, \varinjlim F(G/H')).$$

This can be seen as follows: Given $H' \subset H$, the family consisting only of the map $G/H' \to G/H$ is a covering in T_G. From the associated exact diagram

$$F(G/H) \to F(G/H') \rightrightarrows F(G/H' \times_{G/H} G/H')$$

we see that the map $F(G/H) \to F(G/H')$ identifies the set $F(G/H)$ with the subset $F(G/H')^{H/H'}$ of H/H'-invariant elements in $F(G/H')$. Therefore the map $F(G/H) \to \varinjlim F(G/H')$ identifies the set $F(G/H)$ with the subset $(\varinjlim F(G/H'))^H$ of H-invariant elements in $\varinjlim F(G/H')$. Hence the above map is in fact an isomorphism.

Putting things together, we obtain the canonical isomorphisms

$$\begin{aligned} F(U) &= \varprojlim F(U^H) \\ &\cong \varprojlim \mathrm{Hom}_{G/H}(U^H, F(G/H)) \\ &\cong \varprojlim \mathrm{Hom}_G(U^H, \varinjlim F(G/H')) \\ &\cong \mathrm{Hom}_G(\varinjlim U^H, \varinjlim F(G/H')) \\ &\cong \mathrm{Hom}_G(U, \varinjlim F(G/H')), \end{aligned}$$

which are functorial both in U and in F. Proposition 1.3.3.1 is therefore proved. □

1.3.3.1 implies:

(1.3.3.2) Corollary. *The category of continuous G-modules is equivalent to the category of abelian sheaves on the canonical topology T_G. The equivalence is provided by the mutually quasi-inverse functors $A \mapsto \mathrm{Hom}_G(\,\cdot\,, A)$ and $F \mapsto \varinjlim F(G/H)$.*

The two examples 1.3.2 and 1.3.3 will occur from time to time in the following. The example of the étale topology on a scheme, which is of course the most prominent one in these notes, will be studied from Chapter II on.

§2. Abelian Presheaves on Topologies

(2.1) The Category of Abelian Presheaves

Let T be a topology and let $\mathcal{P}$ be the category of abelian presheaves on T. Then the following holds:

(2.1.1) Proposition. *i) $\mathcal{P}$ is an abelian category satisfying property Ab5) (cp. 0.1.4), and $\mathcal{P}$ has generators (cp. 0.1.4).*
ii) A sequence $F' \to F \to F''$ of abelian presheaves on T is exact in $\mathcal{P}$ if and only if the sequence $F'(U) \to F(U) \to F''(U)$ of abelian groups is exact for all $U \in T$.

This implies (cp. 0.1.4.2):

(2.1.2) Corollary. *The abelian category $\mathcal{P}$ has sufficiently many injective objects.*

Proof of 2.1.1: We apply propositions 0.1.3.1 and 0.1.4.3 to the categories $\mathcal{C} = T^0$ and $\mathcal{C}' = \mathcal{A}b$, where T^0 denotes the dual category of T. □

(2.1.3) Remark. For later use we explicitly describe a family of generators for the category $\mathcal{P}$: Since $\mathbb{Z}$ is a generator for the category $\mathcal{A}b$, the proof of 0.1.4.3 implies that the family $(Z_U)_{U \in T}$ of abelian presheaves, defined by

$$Z_U(V) = \bigoplus_{\mathrm{Hom}(V,U)} \mathbb{Z}, \quad V \in T,$$

is a family of generators for $\mathcal{P}$. For each abelian presheaf F we have an isomorphism

$$F(U) = \mathrm{Hom}(\mathbb{Z}, F(U)) \cong \mathrm{Hom}(Z_U, F),$$

which is functorial in F. In other words: Z_U represents the section functor $\Gamma_U : F \mapsto F(U)$ from $\mathcal{P}$ to $\mathcal{A}b$.

Let $\mathcal{I}$ be a category and let $\mathcal{H}om(\mathcal{I}, \mathcal{P})$ denote the category of functors $F : \mathcal{I} \to \mathcal{P}$. Instead of $F(i)$, $i \in \mathcal{I}$, we also write F_i, and instead of $\varinjlim F$ (cp. 0.3.1) we also write $\varinjlim F_i$.

(2.1.4) Proposition. *i) For each functor $F : \mathcal{I} \to \mathcal{P}$ the inductive limit $\varinjlim F_i$ exists, and we have $(\varinjlim F_i)(U) = \varinjlim F_i(U)$. The functor $\varinjlim : \mathcal{H}om(\mathcal{I}, \mathcal{P}) \to \mathcal{P}$ is additive and left exact.*

ii) If $\mathcal{I}$ is pseudofiltered (cp. 0.3.2), the functor $\varinjlim : \mathcal{H}om(\mathcal{I}, \mathcal{P}) \to \mathcal{P}$ is exact.

Proof: Let $\varinjlim F_i(U)$ denote the inductive limit of the composite functor $\mathcal{I} \xrightarrow{F} \mathcal{P} \xrightarrow{\Gamma_U} \mathcal{A}b$. Then it is immediately clear, that the abelian presheaf $U \mapsto \varinjlim F_i(U)$ on T together with the morphisms of presheaves $F_i(U) \to \varinjlim F_i(U)$ represents the functor $\varinjlim F$ from $\mathcal{P}$ to $\mathcal{E}ns$. The rest now follows from 0.3.1.1 and 0.3.2.1. □

(2.2) Čech-Cohomology

Since the category $\mathcal{P}$ of abelian presheaves on T is an abelian category with sufficiently many injective objects, the methods of homological algebra (cp. 0.2.2.1) imply that the right derived functors exist for each left exact functor from $\mathcal{P}$ into an abelian category.

The section functor $\Gamma_U : \mathcal{P} \to \mathcal{A}b$, defined by $F \mapsto F(U)$, is exact by 2.1.1, and hence $R^q\Gamma_U = 0$ for $q \geq 1$.

Let $\{U_i \to U\}$ be a covering in T. We consider the functor

$$H^0(\{U_i \to U\}, \cdot\,) : \mathcal{P} \to \mathcal{A}b,$$

which assigns to each abelian presheaf F on T the abelian group

$$H^0(\{U_i \to U\}, F) = \ker(\prod_i F(U_i) \rightrightarrows \prod_{i,j} F(U_i \times_U U_j)).$$

This functor is additive and left exact (cp. 2.1.1), but in general not exact.

(2.2.1) Definition. *For an abelian presheaf F on T the* **q-th Čech cohomology group with values in F, associated with** $\{U_i \to U\}$ *is defined as*

$$H^q(\{U_i \to U\}, F) = R^q H^0(\{U_i \to U\}, \ \cdot\ )(F).$$

(2.2.2) Remark. For an **abelian sheaf** F on T the diagram

$$F(U) \to \prod_i F(U_i) \rightrightarrows \prod_{i,j} F(U_i \times_U U_j)$$

is exact, hence $H^0(\{U_i \to U\}, F)$ gets identified with $F(U)$. Therefore, if we consider the section functor $\Gamma_U : \mathcal{S} \to \mathcal{A}b$ on the category of abelian sheaves, it factorizes as

$$\mathcal{S} \xrightarrow{i} \mathcal{P} \xrightarrow{H^0(\{U_i \to U\}, \ \cdot\ )} \mathcal{A}b,$$

where i denotes the inclusion of $\mathcal{S}$ in $\mathcal{P}$. The section functor $\Gamma_U : \mathcal{S} \to \mathcal{A}b$ is left exact (cp. 3.3), but in general not exact. Its right derived functors $H^q(U, \ \cdot\ )$ are the cohomology groups with values in abelian sheaves (cp. 3.3). Let us assume that the conditions of theorem 0.2.3.5 for the existence of spectral sequences attached to the composition of left exact functors are satisfied. In our situation this means that $\mathcal{S}$ is an abelian category with sufficiently many injective objects (cp. 3.2), and that i is left exact and maps injective objects in $\mathcal{S}$ to objects in $\mathcal{P}$, which are $H^0(\{U_i \to U\}, \ \cdot\ )$ - acyclic. We obtain for each abelian sheaf F the spectral sequence

$$E_2^{pq} = H^p(\{U_i \to U\}, R^q i(F)) \Longrightarrow H^{p+q}(U, F),$$

which is functorial in F. This spectral sequence describes the relation between Čech cohomology and cohomology with values in abelian sheaves. More details can be found in 3.4.

The Čech cohomology groups $H^q(\{U_i \to U\}, F)$ can be determined by means of the complex of Čech cochains:

Given a covering $\{U_i \to U\}_{i \in I}$ in T and an abelian presheaf F, we define for $q \geq 0$ the group of q-cochains with values in F, belonging to the covering $\{U_i \to U\}$, by

$$C^q(\{U_i \to U\}, F) = \prod_{(i_0, \dots, i_q) \in I^{q+1}} F(U_{i_0} \times_U \cdots \times_U U_{i_q}).$$

The coboundary operator $d^q : C^q(\{U_i \to U\}, F) \to C^{q+1}(\{U_i \to U\}, F)$ is defined by

$$(d^q s)_{i_0,\dots,i_{q+1}} = \sum_{\nu=0}^{q+1} (-1)^\nu F(\hat{\nu})(s_{i_0,\dots,\widehat{i_\nu},\dots,i_{q+1}}).$$

Here $\hat{\nu} : U_{i_0} \times_U \cdots \times_U U_{i_{q+1}} \to U_{i_0} \times_U \cdots \times_U \widehat{U}_{i_\nu} \times_U \cdots \times_U U_{i_{q+1}}$ denotes the natural projection onto the product obtained by deleting the factor U_{i_ν}. As always, we have $d^{q+1} \circ d^q = 0$ for $q \geq 0$, so that $C^*(\{U_i \to U\}, F)$ is a complex.

(2.2.3) Theorem. *For each abelian presheaf F on T the group $H^q(\{U_i \to U\}, F\}$ can be canonically identified with the q-th cohomology group of the complex $C^*(\{U_i \to U\}, F)$.*

Proof: Let us temporarily denote the cohomology groups of the complex $C^*(\{U_i \to U\}, F)$ by $\tilde{H}^q(\{U_i \to U\}, F)$. This means

$$\tilde{H}^q(\{U_i \to U\}, F) = \ker(d^q)/\mathrm{im}(d^{q-1}) \quad \text{for } q \geq 1,$$
$$\tilde{H}^0(\{U_i \to U\}, F) = \ker(d^0).$$

A morphism $F \to G$ of abelian presheaves induces naturally a homomorphism of complexes $C^*(\{U_i \to U\}, F) \to C^*(\{U_i \to U\}, G)$ and therefore for all $q \geq 0$ a homomorphism $\tilde{H}^q(\{U_i \to U\}, F) \to \tilde{H}^q(\{U_i \to U\}, G)$. Hence we obtain for each $q \geq 0$ a functor

$$\tilde{H}^q(\{U_i \to U\}, \cdot\,) : \mathcal{P} \to \mathcal{A}b,$$

which is obviously additive.

The system of all these functors forms a **∂-functor** from $\mathcal{P}$ to $\mathcal{A}b$ (cp. 0.2.1) in the following manner: If

$$0 \to F' \to F \to F'' \to 0$$

is an exact sequence in $\mathcal{P}$, we obtain a sequence of complexes

$$0 \to C^*(\{U_i \to U\}, F') \to C^*(U_i \to U\}, F) \to C^*(\{U_i \to U\}, F'') \to 0,$$

which is exact by 2.1.1. As is well known, this defines the connecting homomorphisms in cohomology

$$\partial : \tilde{H}^q(\{U_i \to U\}, F) \to \tilde{H}^{q+1}(\{U_i \to U\}, F)$$

for $q \geq 0$. These satisfy the following properties:

i) Given a commutative diagram

$$\begin{array}{ccccccccc} 0 & \longrightarrow & F' & \longrightarrow & F & \longrightarrow & F'' & \longrightarrow & 0 \\ & & \downarrow & & \downarrow & & \downarrow & & \\ 0 & \longrightarrow & G' & \longrightarrow & G & \longrightarrow & G'' & \longrightarrow & 0 \end{array}$$

of exact sequences of morphisms of abelian presheaves, the diagram

$$\begin{array}{ccc} \check{H}^q(\{U_i \to U\}, F'') & \xrightarrow{\partial} & \check{H}^{q+1}(\{U_i \to U\}, F') \\ \downarrow & & \downarrow \\ \check{H}^q(\{U_i \to U\}, G'') & \xrightarrow{\partial} & \check{H}^{q+1}((\{U_i \to U\}, G'). \end{array}$$

commutes.

ii) The long sequence

$$\cdots \to \check{H}^q(\{U_i \to U\}, F'')) \xrightarrow{\partial} \check{H}^{q+1}(\{U_i \to U\}, F')$$
$$\to \check{H}^{q+1}(\{U_i \to U\}, F) \to \ldots,$$

attached to a short exact sequence $0 \to F' \to F \to F'' \to 0$ in $\mathcal{P}$, is exact, hence a forteriori a complex.

More precisely then, the family of functors $\check{H}^q(\{U_i \to U\}, \cdot\,)$ is an **exact** ∂-functor from $\mathcal{P}$ to $\mathcal{A}b$. The family extends the functor $H^0(\{U_i \to U\}, \cdot\,)$, hence we have $\check{H}^0(\{U_i \to U\}, \cdot\,) = H^0(\{U_i \to U\}, \cdot\,)$. In view of 0.2.2 we can finish the proof of the theorem, provided we show that the family $\left(\check{H}^q(\{U_i \to U\}, \cdot\,)\right)_{q \geq 0}$ is **universal**. This is the crucial part of the proof.

By 0.2.1.2 the universality of $\left(\check{H}^q(\{U_i \to U\}, \cdot\,)\right)_{q \geq 0}$ will follow, once we show that the functors $\check{H}^q(\{U_i \to U\}, \cdot\,)$ are effaceable for $q > 0$, i.e. annihilate injective objects from $\mathcal{P}$ (cp. 0.2.1.1). Therefore we have to show that the cochain complex

$$C^0(\{U_i \to U\}, F) \xrightarrow{d^0} C^1(\{U_i \to U\}, F) \xrightarrow{d^1} \cdots$$

is exact for all injective objects in $\mathcal{P}$.

To show this, we use the abelian presheaves Z_U for $U \in T$ introduced in 2.1.3. Recall that

$$Z_U(V) = \bigoplus_{\mathrm{Hom}(V,U)} \mathbb{Z} \quad \text{for} \quad V \in T,$$

and that $\mathrm{Hom}(Z_U, F) \cong F(U)$ for F in $\mathcal{P}$. We therefore obtain the following description for the group $C^q(\{U_i \to U\}, F)$ of q-cochains:

$$\begin{aligned} C^q(\{U_i \to U\}, F) &= \prod_{(i_0,\dots,i_q)} F(U_{i_0} \times_U \dots \times_U U_{i_q}) \\ &\cong \prod_{(i_0,\dots,i_q)} \mathrm{Hom}\,(Z_{U_{i_0} \times_U \dots \times_U U_{i_q}}, F) \\ &\cong \mathrm{Hom}\,(\bigoplus_{(i_0,\dots,i_q)} Z_{U_{i_0} \times_U \dots \times_U U_{i_q}}, F). \end{aligned}$$

Since we assume F to be injective in $\mathcal{P}$, in order to prove the exactness of the cochain complex, it suffices to show exactness of the sequence

$$\bigoplus_i Z_{U_i} \leftarrow \bigoplus_{i,j} Z_{U_i \times_U U_j} \leftarrow \cdots,$$

or – equivalently – of the sequence

$$(*) \qquad \bigoplus Z_{U_i}(V) \leftarrow \bigoplus_{i,j} Z_{U_i \times_U U_j}(V) \leftarrow \cdots$$

for all $V \in T$. Here the arrows are the maps induced by the coboundary operators d^q. Now, $Z_{U_i}(V) = \bigoplus_{\mathrm{Hom}(V,U_i)} \mathbb{Z}$ etc., and it is easy to see that the arrows in $(*)$ are induced by the following diagram of maps of sets:

$$\coprod_i \mathrm{Hom}\,(V, U_i) \leftleftarrows \coprod_{i,j} \mathrm{Hom}\,(V_i, U_i \times_U U_j) \overset{\leftarrow}{\leftleftarrows} \cdots$$

Given $\varphi \in \mathrm{Hom}\,(V, U)$, let $\mathrm{Hom}_\varphi(V, U_i)$ denote the set of all morphisms $V \to U_i$, such that the diagram

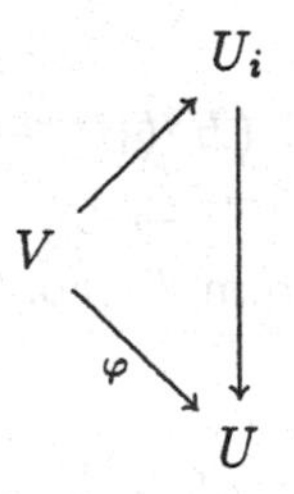

commutes. For the set $\mathrm{Hom}(V, U_i)$ we then obtain the partitions

$$\begin{aligned} \mathrm{Hom}(V, U_i) &= \textstyle\coprod_{\varphi \in \mathrm{Hom}(V,U)} \mathrm{Hom}_\varphi(V, U_i), \\ \mathrm{Hom}(V, U_i \times_U U_j) &= \textstyle\coprod_{\varphi \in \mathrm{Hom}(V,U)} \mathrm{Hom}_\varphi(V, U_i \times_U U_j) \\ &= \textstyle\coprod_{\varphi \in \mathrm{Hom}(V,U)} \mathrm{Hom}_\varphi(V, U_i) \times \mathrm{Hom}_\varphi(V, U_j) \end{aligned}$$

etc.

Let us temporarily abbreviate $\coprod_i \mathrm{Hom}_\varphi(V, U_i)$ by $S(\varphi)$ for a given $\varphi \in \mathrm{Hom}(U, V)$. The diagram above can be rewritten as

$$\coprod_{\varphi \in \mathrm{Hom}(V,U)} [\, S(\varphi) \leftleftarrows S(\varphi) \times S(\varphi) \;\overleftarrow{\overleftarrow{\leftarrow}}\; \cdots].$$

We see that this reduces the proof of the exactness of the sequence $(*)$ to proving the exactness of a sequence of type

$$\bigoplus_S \mathbb{Z} \xrightarrow{d} \bigoplus_{S \times S} \mathbb{Z} \xrightarrow{d} \cdots$$

where the maps are induced by the diagram $S \leftleftarrows S \times S \;\overleftarrow{\overleftarrow{\leftarrow}}\; \cdots$. This means in explicit form that

$$d(1_{i_0,\ldots,i_q}) = \sum_{\nu=0}^{q} (-1)^\nu 1_{i_0,\ldots,\widehat{i_\nu},\ldots,i_q}$$

for basis elements $1_{i_0,\ldots,i_q} \in \bigoplus_{S^{q+1}} \mathbb{Z}$.

Take a fixed $s \in S$. We can define a homomorphism

$$\Delta : \bigoplus_{S^q} \mathbb{Z} \to \bigoplus_{S^{q+1}} \mathbb{Z}$$

by $\Delta(1_{i_0,\ldots,i_q}) = 1_{s,i_0,\ldots,i_q}$. Then $d\Delta + \Delta d = \mathrm{id}$, which shows the exactness of the sequence $\bigoplus_S \mathbb{Z} \leftarrow \bigoplus_{S\times S} \mathbb{Z} \leftarrow \cdots$. Therefore the functor $\big(H^q(\{U_i \to U\}, \,\cdot\,)\big)_{q \geq 0}$ is universal, and theorem 2.2.3 is proved. □

Let $U \in T$ be fixed, and let us consider all the coverings of U in T. We define:

(2.2.4) Definition. *A* **refinement map of coverings of** U

$$\{U'_j \to U\}_{j \in J} \to \{U_i \to U\}_{i \in I}$$

consists of a map $\varepsilon : J \to I$ *of the index sets and of a family* $(f_j)_{j \in J}$ *of* U*-morpisms* $f_j : U'_j \to U_{\varepsilon(j)}$.

The collection of all coverings of U with the refinement maps as morphisms forms a category $\mathcal{J}_U$, called the **category of coverings of** U in the topology T.

Let F be an abelian presheaf on T. For each covering $\{U_i \to U\}$ of U the Čech cohomology groups $H^q(\{U_i \to U\}, F)$ are defined. Let

$f : \{U'_j \to U\} \to \{U_i \to U\}$ be a refinement map. We obtain the canonical homomorphism

$$H^q(f,F) : H^q(\{U_i \to U\},F) \to H^q(\{U'_j \to U\},F),$$

which is ∂-functorial in F, in the following way: From the commutative diagram

$$\begin{array}{ccc} \prod_i F(U_i) & \rightrightarrows & \prod_{i_0,i_1} F(U_{i_0} \times_U U_{i_1}) \\ \downarrow f^0 & & \downarrow f^1 \\ \prod_j F(U'_j) & \rightrightarrows & \prod_{j_0,j_1} F(U'_{j_0} \times_U U'_{j_1}) \end{array}$$

where the vertical arrows f^0 and f^1 are naturally induced by f, we obtain a homomorphism

$$H^0(F,f) : H^0(\{U_i \to U\},F) \to H^0(\{U'_j \to U\},F),$$

which is functorial in F. The universality of the ∂-functor $(H^q(\{U_i \to U\}, \cdot\,))_{q\geq 0}$ implies, that this homomorphism extends uniquely to a morphism $(H^q(\{U_i \to U\}, \cdot\,))_{q\geq 0} \to (H^q(\{U'_j \to U\}, \cdot\,))_{q\geq 0}$ of ∂-functors.

Therefore we obtain for each abelian presheaf F on T and each $q \geq 0$ a contravariant functor

$$\{U_i \to U\} \to H^q(\{U_i \to U\},F)$$

from the category $\mathcal{J}_U$ of coverings of U to the category $\mathcal{A}b$. If we view this as a covariant functor on the dual category of $\mathcal{J}_U$, we can define (cp. 0.3.1.1):

(2.2.5) Definition. *For $q \geq 0$ the group*

$$\check{H}^q(U,F) = \varinjlim_{\{U_i \to U\}} H^q(\{U_i \to U\},F)$$

is the **q-th Čech cohomology group of U with values in F**.

(2.2.6) Theorem. *The functor $F \to \check{H}^0(U,F)$ from the category $\mathcal{P}$ of abelian presheaves to the category $\mathcal{A}b$ is left exact and additive. The right derivatives of this functor are given by the Čech cohomology groups.*

Proof: The crucial point of the proof consists in showing that the limit functor $\varinjlim : \mathcal{H}om(\mathcal{J}_U^0, \mathcal{A}b) \to \mathcal{A}b$ maps exact sequences of functors of the form $\{U_i \to U\} \to H^q(\{U_i \to U\}, F)$ to exact sequences. This is not at all obvious, since the category $\mathcal{J}_U^0$ need not be pseudofiltered (cp. 0.3.2), and therefore the functor $\varinjlim : \mathcal{H}om(\mathcal{J}_U^0, \mathcal{A}b) \to \mathcal{A}b$ may not be exact.

If we assume the exactness property mentioned above, the statements of the theorem follow routinely:

Given an exact sequence $0 \to F' \to F \to F'' \to 0$ in $\mathcal{P}$, we obtain for each covering $\{U_i \to U\}$ the exact sequence

$$0 \to H^0(\{U_i \to U\}, F') \to H^0(\{U_i \to U\}, F) \to H^0(\{U_i \to U\}, F''),$$

which is functorial in $\{U_i \to U\}$. Applying $\varinjlim$, this yields the exact sequence

$$0 \to \check{H}^0(U, F') \to \check{H}^0(U, F) \to \check{H}^0(U, F'').$$

Therefore $F \to \check{H}^0(U, F)$ is left exact.

Moreover, the family of functors $\check{H}^q(U, \cdot\,) : \mathcal{P} \to \mathcal{A}b$ is naturally a ∂-functor, since the homomorphisms

$$H^q(f, F) : H^q(\{U_i \to U\}, F) \to H^q(\{U'_j \to U\}, F)$$

for a refinement map $f : \{U'_j \to U\} \to \{U_i \to U\}$ are ∂-functorial. This ∂-functor is exact, since all cohomology sequences

$$\cdots \to H^q(\{U_i \to U\}, F'') \xrightarrow{\partial} H^{q+1}(\{U_i \to U\}, F') \to \cdots,$$

attached to a given exact sequence $0 \to F' \to F \to F'' \to 0$ in $\mathcal{P}$, are exact, and this remains true after applying $\varinjlim$.

Finally, this ∂-functor is universal, since

$$\check{H}^q(U, F) = \varinjlim H^q(\{U_i \to U\}, F) = 0$$

for $q \geq 1$ and an injective object F from $\mathcal{P}$. From 0.2.2 we conclude, that the right derived functors of $\check{H}^0(U, \cdot\,)$ are given by $\check{H}^q(U, \cdot\,)$.

Let us turn now to the proof of the exactness property mentioned at the beginning. It is based on the following

(2.2.7) Lemma. *Let $\{U_i \to U\}$ and $\{U'_j \to U\}$ be coverings of U, and let*

$$f, g : \{U'_j \to U\} \to \{U_i \to U\}$$

be two refinement maps. Then the induced homomorphisms

$$H^q(f,F), H^q(g,F) : H^q(\{U_i \to U\}, F) \to H^q(\{U'_j \to U\}, F)$$

coincide for all abelian presheaves F and all $q \geq 0$.

To prove the lemma, it suffices to show that $H^0(f,F)$ and $H^0(g,F)$ coincide. Consider the diagram

$$\begin{array}{ccc} \prod F(U_i) & \xrightarrow{d^0} & \prod F(U_{i_0} \times_U U_{i_1}) \\ {\scriptstyle f^0}\downarrow{\scriptstyle g^0} & \overset{\Delta^1}{\swarrow} & {\scriptstyle f^1}\downarrow{\scriptstyle g^1} \\ \prod F(U'_j) & \xrightarrow[d^0]{} & \prod F(U'_{j_0} \times_U U'_{j_1}), \end{array}$$

where $\Delta^1 : \prod F(U_{i_0} \times_U U_{i_1}) \to \prod F(U'_j)$ is defined by $(\Delta^1 s)_j = F((f_j, g_j)_U(s_{\varepsilon(j),\delta(j)}))$. Here $f = (\varepsilon, f_j)$, $g = (\delta, g_j)$, and $(f_j, g_j)_U$ denotes the canonical morphism $U'_j \to U_{\varepsilon(j)} \times_U U_{\delta(j)}$. An easy calculation shows

$$\Delta^1 \circ d^0 = g^0 - f^0.$$

This implies $g^0/\ker d^0 = f^0/\ker d^0$, and therefore $H^0(f,F) = H^0(g,F)$. The lemma is proved. □

Given coverings $\{U'_j \to U\}$ and $\{U_i \to U\}$ of U, we call $\{U'_j \to U\}$ **finer** than $\{U_i \to U\}$, and we denote this by $\{U'_j \to U\} \geq \{U_i \to U\}$, if there is at least one refinement map $\{U'_j \to U\} \to \{U_i \to U\}$. With respect to $\geq$ the collection of all coverings of U forms an (increasingly) filtered set: In fact, given coverings $\{U_i \to U\}_{i \in I}$ and $\{U'_j \to U\}_{j \in J}$, the axioms T1) and T2) for coverings stated in 1.2.1 imply, that

$$\{U_i \times_U U'_j \to U\}_{(i,j) \in I \times J}$$

is also a covering of U, and we have the obvious refinement maps

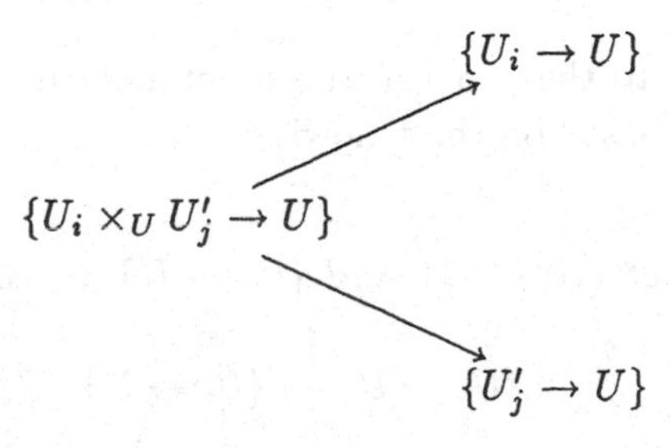

The lemma now tells us, that the functor $\{U_i \to U\} \mapsto H^0(\{U_i \to U\}, F)$ from the category $\mathcal{J}_U^0$ to $\mathcal{A}b$ factors through the ordered set of all coverings, which we view in an obvious way as a category. It is clear that $\varinjlim_{\{U_i \to U\}} H^q(\{U_i \to U\}, F)$ gets identified with the inductive limit over the ordered set of coverings. Since this set is filtered, the last step is exact (cp. 0.3.2.1). Thus, in fact, exact sequences of functors of the form $\{U_i \to U\} \mapsto H^q(\{U_i \to U\}, F)$ remain exact after the application of $\varinjlim$. This finishes the proof of Theorem 2.2.6.

(2.2.8) Remark. For an abelian **sheaf** F on T we have

$$\check{H}^0(U, F) = \varinjlim_{\{U_i \to U\}} H^0(\{U_i \to U\}, F) = \varinjlim_{\{U_i \to U\}} F(U) = F(U).$$

Therefore, the section functor Γ_U factors on the category $\mathcal{S}$ of abelian sheaves as

$$\Gamma_U = \check{H}^0(U, \,\cdot\,) \circ i,$$

where $i : \mathcal{S} \to \mathcal{P}$ is the inclusion.

(2.3) The Functors f^p and f_p

Let T and T' be topologies, and let $f : T \to T'$ be a functor of the underlying categories. We do not assume that f is a morphism of topologies, although this is the most interesting case.

Let $\mathcal{P}$ and $\mathcal{P}'$ denote the categories of abelian presheaves on T and T' respectively.

Given an abelian presheaf F' on T', we obtain an abelian presheaf f^pF' on T by $f^pF'(U) = F'(f(U))$ for U in T. Moreover, for each morphism $v' : F' \to G'$ of presheaves on T', we obtain a morphism $f^pv' : f^pF' \to f^pG'$ defined by $f^pv'(U) = v'(f^p(U))$ for $U \in T$. The resulting functor $f^p : \mathcal{P}' \to \mathcal{P}$ is additive and **exact** and commutes with inductive limits (cp. 2.1.4).

(2.3.1) Theorem. *i) The functor f^p has a left adjoint functor $f_p : \mathcal{P} \to \mathcal{P}'$, which is right exact, additive and commutes with inductive limits.*

ii) If $f_p : \mathcal{P} \to \mathcal{P}'$ is even exact, then $f^p : \mathcal{P}' \to \mathcal{P}$ maps injective objects in $\mathcal{P}'$ to injective objects in $\mathcal{P}$.

Proof: Let us show ii) first, assuming i). Let M' be an injective object in $\mathcal{P}'$. Consider an exact sequence

$$0 \to F_1 \to F_2 \to F_3 \to 0$$

in $\mathcal{P}$. Since by assumption f_p is exact, we obtain an exact sequence

$$0 \to f_pF_1 \to f_pF_2 \to f_pF_3 \to 0$$

in $\mathcal{P}'$, and therefore an exact sequence

$$0 \to \mathrm{Hom}(f_pF_1, M') \to \mathrm{Hom}(f_pF_2, M') \to \mathrm{Hom}(f_pF_3, M') \to 0,$$

since M' is injective. Now, f_p is left adjoit to f^p, and therefore the sequence

$$0 \to \mathrm{Hom}(F_1, f^pM') \to \mathrm{Hom}(F_2, f^pM') \to \mathrm{Hom}(F_3, f^pM') \to 0$$

is exact as well, and hence f^pM' is injective.

To prove i) we have to show the following: For all $F \in \mathcal{P}$ the functor $G' \mapsto \mathrm{Hom}(F, f^pG')$ from $\mathcal{P}'$ to $\mathcal{A}b$ is representable. This means that for each $F \in \mathcal{P}$ there exists an object f_pF in $\mathcal{P}'$, and for each $G' \in \mathcal{P}'$ an isomorphism

$$\mathrm{Hom}(f_pF, G') \cong \mathrm{Hom}(F, f^pG')$$

of abelian groups, which is functorial in G'. As we explained in 0.1.1, this makes f_p naturally a functor from $\mathcal{P}$ to $\mathcal{P}'$, which is left adjoint to f^p. The functor f_p is then automatically additive and right exact, and commutes with inductive limits (cp. 0.3.1.3).

Let F be an abelian presheaf on T. First of all, we have to define $f_pF(U')$ for all $U' \in T'$. Consider all pairs (U, ϕ') with U an object in T and ϕ' a morphism $\phi' : U' \to f(U)$. The collection of all these pairs will form a category $\mathcal{I}_{U'}$ if we define a morphism $(U_1, \phi'_1) \to (U_2, \phi'_2)$ to be a morphism $\phi : U_1 \to U_2$ such that the following diagram commutes:

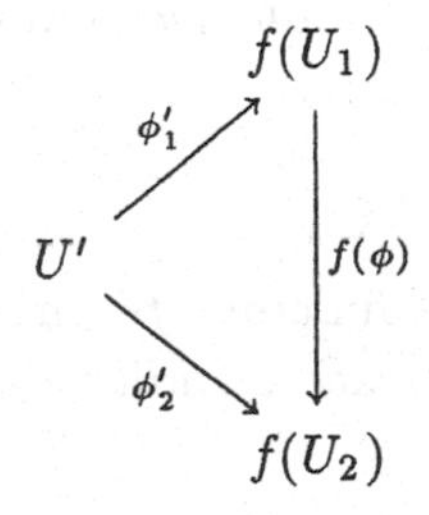

The assignment $(U, \phi') \mapsto F(U)$ leads then to a contravariant functor $F_{U'}$ from the category $\mathcal{I}_{U'}$ of pairs to the category $\mathcal{Ab}$, and we define (cp. 0.3.1):

$$f_p F(U') = \varinjlim_{\mathcal{I}^0_{U'}} F_{U'} = \varinjlim_{(U,\phi')} F(U).$$

A morphism $\varepsilon' : U' \to V'$ in T' induces a functor $\mathcal{I}_{V'} \to \mathcal{I}_{U'}$, if we map the pair $(V, \phi' : V' \to f(V))$ to the pair $(V, \phi' \circ \varepsilon' : U' \to f(V))$. In this way we obviously obtain a homomorphism

$$\varinjlim_{\mathcal{I}^0_{V'}} F_{V'} \to \varinjlim_{\mathcal{I}^0_{U'}} F_{U'},$$

hence a homomorphism $f_p F(V') \to f_p F(U')$.

Therefore, we have constructed an abelian presheaf $f_p F$ on T', and we are left to show the existence of isomorphisms

$$\mathrm{Hom}(f_p F, G') \cong \mathrm{Hom}(F, f^p G'),$$

which are functorial in $G' \in \mathcal{P}'$.

Let $v : f_p F \to G'$ be a given morphism in $\mathcal{P}'$. For each $U \in T$ this yields a homomorphism

$$v(f(u)) : f_p F(f(U)) \to G'(f(U)) = f^p G'(U).$$

Since the pair $(U, \mathrm{id}_{f(U)})$ is an object of the category $\mathcal{I}_{f(U)}$, there is a corresponding canonical homomorphism

$$F(U) = F_{f(U)}(U, \mathrm{id}_{f(U)}) \to \varinjlim_{\mathcal{I}^0_{f(U)}} F_{f(U)} = f_p F(f(U)).$$

The composite of the two homomorphisms yields a homomorphism $F(U) \to f^p G'(U)$, which is clearly functorial in U. We therefore obtain a morphism

$$v^b : F \to f^p G'$$

of abelian presheaves on T, and it is easily checked that the corresponding map

$$\mathrm{Hom}(f_p F, G') \to \mathrm{Hom}(F, f^p G')$$

is a homomorphism of abelian groups, which is functorial in G'.

Conversely, let us assume that $u : F \to f^p G'$ is a given morphism in $\mathcal{P}$. Let $U' \in T'$. We then obtain for each object (U, φ') in $\mathcal{I}_{U'}$ a homomorphism

$$F_{U'}(U, \varphi') = F(U) \overset{u(U)}{\to} f^p G'(U) = G'(f(U)) \overset{G'(\varphi')}{\to} G'(U'),$$

which is functorial in (U, φ'). This induces a homomorphism

$$f_p F(U') = \varinjlim_{\mathcal{I}^0_{U'}} F_{U'} \to G'(U'),$$

which itself is functorial in $U' \in T'$. Thus we have a morphism

$$u^{\#} : f_p F \to G'$$

of abelian presheaves on T', and we find that $u \to u^{\#}$ is inverse to $v \to v^b$. This finishes the proof of Theorem 2.3.1. □

(2.3.2) Remark. Obviously, the functor f^p can also be viewed as a functor from the category of presheaves of **sets** on T' to the category of presheaves of sets on T. The existence of the left adjoint functor f_p in this situation is proved essentially with the same construction as in the proof of 2.3.1 (we use 0.3.1.2). See also [2], exp. I, 5.1.

(2.3.3) Example. We keep the same notations as above. We claim:

If the presheaf F on T is representable by an object $Z \in T$, i.e. $F(U) = \mathrm{Hom}(U, Z)$ for U in T, then the presheaf $f_p F$ is represented by the object $f(Z) \in T'$.

Proof: Let $U' \in T'$. Each $(U, \varphi') \in \mathcal{I}_{U'}$ induces a map

$$F(U) = \mathrm{Hom}(U, Z) \to \mathrm{Hom}(U', f(Z)),$$

by assigning to a morphism $U \overset{\alpha}{\to} Z$ the composite $U' \overset{\varphi'}{\to} f(U) \overset{f(\alpha)}{\to} f(Z)$. Since this map is functorial in (U, φ'), we obtain a canonical map

$$f_p F(U') = \varinjlim_{(U,\varphi')} \mathrm{Hom}(U, Z) \to \mathrm{Hom}(U', f(Z),$$

which is functorial in U'. This map is surjective: Given $\varphi' \in \mathrm{Hom}(U', f(Z))$, the pair (Z, φ') belongs to $\mathcal{I}_{U'}$, and under the map $\mathrm{Hom}(Z, Z) \to \mathrm{Hom}(U', \varphi(Z))$, corresponding to (Z, φ'), the element id_Z is mapped to φ'. The map is also injective: Assume there are (U_1, φ'_1),

(U_2, φ_2') in $\mathcal{I}_{U'}$, and morphisms $\alpha_1 : U_1 \to Z$, $\alpha_2 : U_2 \to Z$, such that $f(\alpha_1)\varphi_1' = f(\alpha_2)\varphi_2'$. Under the morphism $\mathrm{Hom}(Z,Z) \to \mathrm{Hom}(U_1,Z)$ corresponding to $\alpha_1 : (U_1, \varphi_1') \to (Z, f(\alpha_1)\varphi_1')$, the element id_Z is mapped to α_1, and – similarly – the element id_Z is mapped to α_2 under the morphism $\mathrm{Hom}(Z,Z) \to \mathrm{Hom}(U_2,Z)$ corresponding to $\alpha_2 : (U_2, \varphi_2') \to (Z, f(\alpha_2)\varphi_2')$. □

(2.3.4) Example. Let P denote the topology, whose underlying category consists of a single object and a single arrow. The category of abelian presheaves on P gets identified with $\mathcal{Ab}$.

Let T be a topology and U an object in T. There is a unique functor $i : P \to T$, which maps the single object in P to U. We are going to compute the functors $i^p : \mathcal{P} \to \mathcal{Ab}$ and $i_p : \mathcal{Ab} \to \mathcal{P}$.

Trivially, for an abelian presheaf F on T we have

$$i^p(F) = F(U).$$

Conversely, assume A is an abelian group. To compute the presheaf $i_p A$ on T, we consider $V \in T$ and the category $\mathcal{I}_V$. The objects in $\mathcal{I}_V$ get identified with the morphisms $\varphi : V \to U$, and for two morphisms $\varphi, \psi : V \to U$ we have

$$\mathrm{Hom}_{\mathcal{I}_V}(\varphi, \psi) = \begin{cases} \emptyset & \text{if } \varphi \neq \psi \\ \mathrm{id} & \text{if } \varphi = \psi. \end{cases}$$

Therefore, $\mathcal{I}_V$ can be identified with the discrete category on the set $\mathrm{Hom}(V,U)$. This implies

$$i_p A(V) = \bigoplus_{\mathrm{Hom}(V,U)} A.$$

Since the formation of direct sums of abelian groups is an exact functor (cp. 0.3.2.1), we obtain from 2.3.1, ii): If F is an injective abelian presheaf on T, the abelian group $F(U)$ is injective for all $U \in T$.

In particular, if we take $A = \mathbb{Z}$, the presheaf $i_p \mathbb{Z}$ is equal to the presheaf Z_U mentioned in 2.1.3, and the canonical isomorphisms $\mathrm{Hom}(Z_U, F) \cong F(U)$, $F \in \mathcal{P}$, which we considered there, are from our current point of view simply obtained from the fact that i_p is left adjoint to i^p.

§ 3. Abelian Sheaves on Topologies

(3.1) The Associated Sheaf of a Presheaf

Let T be a topology, $\mathcal{P}$ be the category of abelian presheaves on T, and let $\mathcal{S}$ be the category of abelian sheaves on T. $\mathcal{S}$ is a full subcategory of $\mathcal{P}$. Let $i : \mathcal{S} \hookrightarrow \mathcal{P}$ denote the inclusion. The goal of this section is the proof of the following

(3.1.1) Theorem. *The left adjoint functor* $^{ad}i : \mathcal{P} \to \mathcal{S}$ *of* $i : \mathcal{S} \to \mathcal{P}$ *exists.*

Remark. Given a presheaf F on T, we write $F^{\#}$ for the abelian sheaf $^{ad}i(F)$, and call $F^{\#}$ the **sheaf associated to the presheaf** F. The condition for i and ^{ad}i to be adjoint can also be formulated in the following way: There is a functorial morphism $F \to F^{\#}$ (the adjoint morphism $F \to i \circ {}^{ad}i(F)$, cp. (0.1.1)), which is **universal for morphisms from** F **to abelian sheaves.** This means that each morphism from F to an abelian sheaf G factors uniquely as $F \to F^{\#} \to G$.

The proof of Theorem (3.1.1) starts with the construction of a functor $\dagger : \mathcal{P} \to \mathcal{P}$: Given an abelian presheaf F on T, we define

$$F^{\dagger}(U) = \check{H}^0(U, F) = \varinjlim_{\{U_i \to U\}} H^0(\{U_i \to U\}, F)$$

for $U \in T$ (cp. (2.2.5)). If $V \to U$ is a morphism in T, we have a functor $\{U_i \to U\} \to \{U_i \times_U V \to V\}$ from the category of coverings of U to the category of coverings of V. This induces a natural homomorphism $F^{\dagger}(U) \to F^{\dagger}(V)$. Therefore $F^{\dagger}$ is an abelian presheaf. Moreover, any morphism $u : F \to G$ in $\mathcal{P}$ induces in a natural way a morphism $u^{\dagger} : F^{\dagger} \to G^{\dagger}$.

For all abelian presheaves F we have a canonical morphism $F \to F^{\dagger}$, which is functorial in F, and given by

$$F(U) = H^0(\{U \stackrel{id}{\to} U\}, F\} \to \check{H}^0(U, F) = F^{\dagger}(U)$$

for $U \in T$.

In particular, for an abelian sheaf G the canonical morphism $G \to G^{\dagger}$ is an isomorphism, and we conclude:

(3.1.2) Every morphism $F \to G$ from the presheaf F to the abelian sheaf G factors uniquely as $F \to F^\dagger \to G$.

Here the uniqueness of the factorization is seen as follows: If $F \to G$ is the zero morphism, and if $\{U_i \to U\}$ is a covering of $U \in T$, the commutative diagram

$$\begin{array}{ccc} & H^0(\{U_i \to U\}, F) & \subset & \prod F(U_i) \\ & \downarrow & & \downarrow \\ G(U) \overset{\cong}{\to} & H^0(\{U_i \to U\}, G) & \subset & \prod G(U_i) \end{array}$$

implies that $H^0(\{U_i \to U\}, F) \to G(U)$ is the zero map. The same is then true for $\check{H}^0(U, F) \to G(U)$.

The theorem will follow from 3.1.2, provided we can show that in case $f^\dagger$ is not already a sheaf, then at least $(F^\dagger)^\dagger$ will always be a **sheaf** for each abelian presheaf F on T. This is in fact true: Let us call a presheaf **separated** if $F(U) \to \prod F(U_i)$ is injective for all coverings $\{U_i \to U\}$. Then – more precisely – the following holds:

(3.1.3) Proposition. *i) For each abelian presheaf F the presheaf $F^\dagger$ is separated.*

ii) If F is a separated abelian presheaf, then $F \to F^\dagger$ is a monomorphism and $F^\dagger$ is a sheaf.

iii) The functor $\dagger: \mathcal{P} \to \mathcal{P}$ is left exact.

For later purposes we have included the last statement, which was part of Theorem 2.2.6.

Proof of (3.1.3): i) Let $\{U_i \to U\}$ be a covering in T, and let

$$\bar{s} \in ker(F^\dagger(U) \to \prod_i F^\dagger(U_i)).$$

We have to show that $\bar{s} = 0$.

Given $\bar{s} \in F^\dagger(U) = \check{H}^0(U, F)$, there are a covering $\{V_j \to U\}$ and an element $s \in H^0(\{V_j \to U\}, F)$ representing $\bar{s}$. This means that $\bar{s}$ is the image of s under the canonical homomorphism $H^0(\{V_j \to U\}, F) \to \check{H}^0(U, F)$.

Let s_i denote the image of s under the homomorphism

$$H^0(\{V_j \to U\}, F) \to H^0(\{V_j \times_U U_i \to U_i\}, F).$$

Then s_i represents the element $(\bar{s})_i = F^\dagger(\varphi_i)(\bar{s})$ in $F^\dagger(U_i)$, where $\varphi_i : U_i \to U$.

By assumption we have $(\bar{s})_i = 0$. Using the description of $\check{H}^0(U_i, F)$ as an inductive limit over a filtered set (cp. the proof of (2.2.6)), we find a refinement map

$$f_i : \{W_{il} \to U_i\} \to \{V_j \times_U U_i \to U_i\},$$

such that $H^0(f_i, F)(s_i) = 0$. If we compose the coverings $\{W_{il} \to U_i\}$ and $\{U_i \to U\}$, we obtain a covering $\{W_{il} \to U\}$. Moreover, f_i induces in a natural way a refinement map

$$f : \{W_{il} \to U\} \to \{V_j \to U\}.$$

By construction we have $H^0(f, F)(s) = 0$. Hence $\bar{s} = 0$.

ii) Let F be a separated abelian presheaf. As a first step we show the following: The canonical homomorphism

$$H^0(\{U_i \to U\}, F) \to \check{H}^0(U, F)$$

is injective for each covering $\{U_i \to U\}$. In particular, the map $F(U) = H^0(\{U \overset{id}{\to} U\}, F) \to \check{H}^0(U, F)$ is injective, and therefore $F \to F^\dagger$ is a monomorphism.

Since $\check{H}^0(U, F)$ can be described as the inductive limit of the groups $H^0(\{U_i \to U\}, F)$ over the filtered set of coverings of U, it suffices to verify the injectivity of the map

$$H^0(f, F) : H^0(\{U_i \to U\}, F) \to H^0(\{V_j \to U\}, F)$$

for a refinement map $f : \{V_j \to U\} \to \{U_i \to U\}$.

Let us consider the covering $\{V_j \times_U U_i \to U\}$ obtained as the composite of the coverings $\{V_j \times_U U_i \to U_i\}$ and $\{U_i \to U\}$, together with the two refinement maps

$$\{V_j \times_U U_i \to U\} \overset{pr_2}{\to} \{U_i \to U\}$$

and

$$\{V_j \times_U U_i \to U\} \overset{pr_1}{\to} \{V_j \to U\} \overset{f}{\to} \{U_i \to U\}.$$

By (2.2.7) we have

$$H^0(pr_2, F) = H^0(f \circ pr_1, F) = H^0(pr_1, F) \circ H^0(f, F),$$

and hence it suffices to show that $H^0(pr_2, F)$ is injective. This map is obtained from

$$\prod_i F(U_i) \to \prod_{j,i} F(V_j \times_U U_i)$$

by restriction to $H^0(\{U_i \to U\}, F)$. Since F is separated, the map $F(U_i) \to \prod_j F(V_j \times_U U_i)$ is injective for each covering $\{V_j \times_U U_i \to U_i\}$, and hence $\prod_i F(U_i) \to \prod_{i,j} F(V_j \times_U U_i)$ is injective as well.

We show next that $F^\dagger$ is a sheaf. This means that the diagram

$$F^\dagger(U) \to \prod_i F^\dagger(U_i) \rightrightarrows \prod_{i,j} F^\dagger(U_i \times_U U_j)$$

is exact for each covering $\{U_i \to U\}$ in T. Since $F^\dagger$ is separated by i), the first map is injective. Let $\bar{s} \in ker(\prod_i F^\dagger(U_i) \rightrightarrows \prod_{i,j} F^\dagger(U_i \times_U U_j)$ and let $\bar{s}_i \in F^\dagger(U_i)$ denote the i-th component of $\bar{s}$. For each i we choose a covering $\{V_{i\nu} \to U_i\}$ and an element $s_i \in H^0(\{V_{i\nu} \to U_i\}, F)$ representing $\bar{s}_i$. Let s^1_{ij} denote the image of s_i under the map

$$H^0(\{V_{i\nu} \to U_i\}, F) \to H^0(\{V_{i\nu} \times_U U_j \to U_i \times_U U_j\}, F)$$

and, similarly, let s^2_{ij} denote the image of s_j under the map

$$H^0(\{V_{j\mu} \to U_j\}, F) \to H^0(\{U_i \times_U V_{j\mu} \to U_i \times_U U_j\}, F).$$

The elements in $F^\dagger(U_i \times_U U_j)$ represented by s^1_{ij} and s^2_{ij} are the images of $\bar{s}_i$ under $F^\dagger(U_i) \to F^\dagger(U_i \times_U U_j)$ and of $\bar{s}_j$ under $F^\dagger(U_j) \to F^\dagger(U_i \times_U U_j)$, hence are the same by our assumption on $\bar{s}$. By what we proved above we obtain "$s^1_{ij} = s^2_{ij}$" on each common refinement of the coverings $\{V_{i\nu} \times_U U_j \to U_i \times_U U_j\}$ and $\{U_i \times_U V_{i\mu} \to U_i \times_U U_j\}$, in particular on $\{V_{i\nu} \times_U V_{j\mu} \to U_i \times_U U_j\}$. Hence we have "$s^1_{ij} = s^2_{ij}$" in $H^0(\{V_{i\nu} \times_U V_{j\mu} \to U_i \times_U U_j\}, F) \subset \prod_{\nu,\mu} F(V_{i\nu} \times_U V_{j\mu})$. But this means that

$$s' := (s_i)_i \in ker(\prod_{i,\nu} F(V_{i\nu}) \rightrightarrows \prod_{i,\nu,j,\mu} F(V_{i\nu} \times_U V_{j\mu})) = H^0(\{V_{i\nu} \to U\}, F),$$

and therefore $\bar{s}' \in F^\dagger(U)$ is mapped to $\bar{s}$ under $F^\dagger(U) \to \prod_i F^\dagger(U_i)$. This proves ii). □

(3.1.4) Corollary. *For an abelian presheaf F on T the following are equivalent:*

i) F is a sheaf.

ii) For each covering $\{U_i \to U\}$ in T there exists a refinement $\{U'_j \to U\}$ of $\{U_i \to U\}$ in T, such that

$$F(U) \to \prod F(U'_j) \rightrightarrows \prod F(U'_{j_0} \times_U U'_{j_1})$$

is exact.

Proof: i) $\Longrightarrow$ ii) holds trivially. ii) $\Longrightarrow$ i) can be seen as follows: Let $\mathcal{J}_U$ denote the category of coverings of U and let $\mathcal{J}'_U$ denote the full subcategory of all coverings $\{U'_j \to U\}$, such that $F(U) \to \prod F(U'_j) \rightrightarrows \prod F(U'_{j_0} \times_U U'_{j_1})$ is exact. Assumption ii) says that $(\mathcal{J}'_U)^0$ is a final subcategory (cp. (0.3.3)) of $\mathcal{J}^0_U$. Therefore (cp. (0.3.3.1))

$$F^\dagger(U) = \varinjlim_{\mathcal{J}^0_U} H^0(\{U_i \to U\}, F) = \varinjlim_{(\mathcal{J}'_U)^0} H^0(\{U'_j \to U\}, F) \xrightarrow{\cong} F(U),$$

hence $F \to F^\dagger$ is an isomorphism, and (3.1.3) implies that F is a sheaf.

(3.1.5) Remark. Essentially the same methods can be used to prove the analogs of (3.1.1) and (3.1.4) also for presheaves and sheaves of sets. Compare [2], exp. II, 6.4.

(3.2) The Category of Abelian Sheaves

(3.2.1) Theorem. *i) The category $\mathcal{S}$ of abelian sheaves on a topology T is an abelian category, which satisfies axiom Ab 5) and has generators.*

ii) The functor $i : \mathcal{S} \to \mathcal{P}$ is left exact and the functor $\# = {}^{ad}i : \mathcal{P} \to \mathcal{S}$ is exact.

Proof of i): a) Since $\mathcal{S}$ is a full subcategory of $\mathcal{P}$, for each pair $F, G \in \mathcal{S}$ the set $\mathrm{Hom}(F, G)$ has the structure of an abelian group, such that composition of morphisms is bilinear. The sheaf $O(U) = O$ for all $U \in T$ is the zero object of $\mathcal{S}$.

b) Let $F \to G$ be a morphism in $\mathcal{S}$. We claim that the presheaf kernel K of $F \to G$, defined by $K(U) = ker(F(U) \to G(U))$ for $U \in T$, is an abelian sheaf and that K, together with the canonical morphism $K \to F$, is the kernel of $F \to G$ in the category $\mathcal{S}$.

In fact, let $\{U_i \to U\}$ be a covering in T. From the commutative diagram

$$\begin{array}{ccccc} & & 0 & & 0 & & 0 \\ & & \downarrow & & \downarrow & & \downarrow \\ & & K(U) & \to & \prod K(U_i) & \rightrightarrows & \prod K(U_i \times_U U_j) \\ & & \downarrow & & \downarrow & & \downarrow \\ & & F(U) & \to & \prod F(U_i) & \rightrightarrows & \prod F(U_i \times_U U_j) \\ & & \downarrow & & \downarrow & & \downarrow \\ & & G(U) & \to & \prod G(U_i) & \rightrightarrows & \prod G(U_i \times_U U_j) \end{array}$$

with exact columns and exact second and third row, we obtain that the first row is exact as well. Hence K is a sheaf. Since K is the kernel of $F \to G$ in $\mathcal{P}$, we have

$$0 \to \operatorname{Hom}(X, K) \to \operatorname{Hom}(X, F) \to \operatorname{Hom}(X, G)$$

exact for all X in $\mathcal{P}$. In particular, this holds for X in $\mathcal{S}$. But $\mathcal{S}$ is a full subcategory of $\mathcal{P}$, and therefore K is the kernel of $F \to G$ in the category $\mathcal{S}$.

Let $F \to G$ be a morphism of abelian presheaves with cokernel C and image I in the category $\mathcal{P}$. We claim that $C^\#$ together with the morphism $G^\# \to C^\#$, induced by $G \to C$, is the cokernel of $F^\# \to G^\#$ in the category $\mathcal{S}$, and that $I^\#$ together with the morphism $I^\# \to G^\#$, induced by $I \to G$, is the image. In particular we see, that for a morphism $F \to G$ in $\mathcal{S}$, the cokernel is the sheaf associated to the presheaf $U \mapsto coker(F(U) \to G(U))$ and the image is the sheaf associated to the presheaf $U \mapsto im(F(U) \to G(U))$.

The claim is proved as follows: Since $\# : \mathcal{P} \to \mathcal{S}$ is left adjoint to $i : \mathcal{S} \to \mathcal{P}$, we have for each $X \in \mathcal{S}$ the following commutative diagram:

$$\begin{array}{ccccccc} 0 & \longrightarrow & \operatorname{Hom}(C, X) & \longrightarrow & \operatorname{Hom}(G, X) & \longrightarrow & \operatorname{Hom}(F, X) \\ & & \downarrow\cong & & \downarrow\cong & & \downarrow\cong \\ 0 & \longrightarrow & \operatorname{Hom}(C^\#, X) & \longrightarrow & \operatorname{Hom}(G^\#, X) & \longrightarrow & \operatorname{Hom}(F^\#, X) \end{array}$$

Here the first row is exact, and hence also the second, which shows that $C^\#$, together with the induced morphism $G^\# \to C^\#$, is the cokernel of $F^\# \to G^\#$ in the category $\mathcal{S}$.

With respect to the image I of $F \to G$, we have the exact sequence

$$0 \to I \to G \to C \to 0$$

in $\mathcal{P}$. Since the functor $i \circ \# = \dagger \circ \dagger : \mathcal{P} \to \mathcal{P}$ is left exact by (3.1.3) iii), we obtain an exact sequence

$$0 \to I^{\#} \to G^{\#} \to C^{\#}$$

in $\mathcal{P}$. This implies

$$I^{\#} \xrightarrow{\cong} ker(G^{\#} \to C^{\#})$$

in $\mathcal{P}$. But we saw above that the presheaf kernel of a morphism in $\mathcal{S}$ is the same as the kernel in $\mathcal{S}$, and therefore $I^{\#}$ with the induced morphism $I^{\#} \to G^{\#}$ is the image of $F^{\#} \to G^{\#}$ in $\mathcal{S}$.

We come to the proof of property Ab 2) (cp. (0.1.3)) for $\mathcal{S}$. Let $v : F \to G$ be a morphism in $\mathcal{S}$, let J denote the coimage and let I denote the image of $v : F \to G$ in the category $\mathcal{P}$. In $\mathcal{P}$ we have a canonical morphism

$$\bar{v} : J \to I,$$

uniquely determined by the fact that the composite $F \to J \xrightarrow{\bar{v}} I \to G$ is equal to v. Since $\bar{v}$ is an isomorphism, the same is true for $\bar{v}^{\#} : J^{\#} \to I^{\#}$. From above we know that $J^{\#}$ and $I^{\#}$ are the coimage and the image of $F \xrightarrow{v} G$ in $\mathcal{S}$. Applying $\#$ to the factorization $F \to J \xrightarrow{\bar{v}} I \to G$ of V, we see that $\bar{v}^{\#}$ is the canonical morphism $J^{\#} \to I^{\#}$ belonging to v in the category $\mathcal{S}$.

c) Let $(F_i)_{i \in I}$ be a family of abelian sheaves on T. The presheaf $\prod_i F_i$, defined by $(\prod_i F_i)(U) = \prod_i F_i(U)$ is obviously a sheaf. Together with the canonical projections $\prod_i F_i \to F_i$ the sheaf $\prod_i F_i$ is the product of the F_i in the category $\mathcal{S}$.

Let F be the direct sum of the F_i in the category $\mathcal{P}$. This means $F(U) = \bigoplus_i F_i(U)$ for $U \in T$ (cp. the proof of (0.1.4.3)). For each X in $\mathcal{P}$ the map

$$\mathrm{Hom}(F, X) \to \prod_i \mathrm{Hom}(F_i, X),$$

induced by the canonical injections $F_i \to F$, is bijective. In particular, if X is in $\mathcal{S}$, the universal mapping property of $F \to F^{\#}$ implies that the map

$$\mathrm{Hom}(F^{\#}, X) \to \prod_i \mathrm{Hom}(F_i, X)$$

is bijective. Therefore $F^{\#}$ with the induced morphisms $F_i \to F^{\#}$ is the direct sum of the F_i in the category $\mathcal{S}$.

d) We show now that $\mathcal{S}$ has property Ab 5) (cp. (0.1.4)). Let C be an abelian sheaf on T, and let $(A_i)_{i\in I}$ be an increasingly filtered family of abelian subsheaves of C. For all $i \in I$ we assume that a morphism $v_i : A_i \to B$ from A_i to an abelian sheaf B is given, such that v_i is induced by v_j if $A_i \subset A_j$. We have to show that there exists an extension

$$v : A \to B$$

of the v_i to the supremum A of the A_i in $\mathcal{S}$ i.e. $A = im(\bigoplus_i A_i \to C)$. Consider the supremum $\bar{A}$ of the A_i in the category $\mathcal{P}$. From parts b) and c) we see that $\bar{A}^{\#}$ gets canonically identified with the supremum A of the A_i in $\mathcal{S}$. In $\mathcal{P}$ there exists an extension $\bar{v} : \bar{A} \to B$ of the v_i. Since B is an abelian sheaf, the map $\bar{v}$ factorizes uniquely, as

$$\bar{A} \to \bar{A}^{\#} \xrightarrow{\bar{v}^{\#}} B.$$

After identifying $\bar{A}^{\#}$ and A, we obtain in $v = \bar{v}^{\#}$ the extension of the v_i to A.

e) In (2.1.3) we mentioned explicitely a family $(Z_U)_{U\in T}$ of abelian presheaves on T, which generate $\mathcal{P}$. For each $F \in \mathcal{S}$ we have

$$\mathrm{Hom}(Z_U^{\#}, F) \cong \mathrm{Hom}(Z_U, F) \cong F(U).$$

This shows that the family $(Z_U^{\#})_{U\in T}$ is a family of generators for the category $\mathcal{S}$.

Proof of ii): It follows from part b) above that an exact sequence $0 \to K \to F \to G$ in $\mathcal{S}$ is also exact in $\mathcal{P}$. Hence the functor $i : \mathcal{S} \to \mathcal{P}$ is left exact.

The functor $\# : \mathcal{P} \to \mathcal{S}$ is left adjoint to $i : \mathcal{S} \to \mathcal{P}$ and therefore right exact. By (3.1.3), iii) the composite $i \circ \# = \dagger \circ \dagger : \mathcal{P} \to \mathcal{P}$ is left exact. But i is fully faithful, and hence $\#$ is left exact. □

(3.2.2) Corollary. *The category $\mathcal{S}$ of abelian sheaves on T is an abelian category with sufficiently many injective objects. (cp. (0.1.4.2))*

(3.2.3) Corollary. *Let $\mathcal{I}$ be a category.*

i) For each functor $F : \mathcal{I} \to \mathcal{S}$ the inductive limit $\varinjlim F$ exists in $\mathcal{S}$. It is equal to the sheaf associated to the presheaf $U \mapsto \varinjlim F_i(U)$. (cp. (2.1.4)). The functor $\varinjlim : \mathcal{H}om(\mathcal{I}, \mathcal{S}) \to \mathcal{S}$ is right exact.

ii) If the category $\mathcal{I}$ is pseudofiltered, the functor $\varinjlim : \mathcal{H}om(\mathcal{I}, \mathcal{S}) \to \mathcal{S}$ is exact.

Proof: i) follows from (0.3.1.1) and the fact that # commutes with inductive limits (cp. (0.3.1.3)). ii) follows from (0.3.2.1). □

(3.3) Cohomology of Abelian Sheaves

The category $\mathcal{S}$ of abelian sheaves on T is an abelian category with sufficiently many injective objects (cp. (3.2.2)). Therefore the right derived functors $R^q f$ exist for each given left exact additive functor $f : \mathcal{S} \to \mathcal{C}$ from $\mathcal{S}$ into an abelian category $\mathcal{C}$ (cp. (0.2.2.1)).

In this part we consider for a fixed $U \in T$ the section functor

$$\Gamma_U : \mathcal{S} \to \mathcal{A}b$$

defined by $\Gamma_U(F) = F(U)$. This functor is additive and left exact as the composite of the left exact functor $i : \mathcal{S} \to \mathcal{P}$ (cp. (3.2.1) ii)) and the exact section functor on $\mathcal{P}$ (cp. (2.1.1) ii)).

(3.3.1) Definition. *For an abelian sheaf F on T we define* **the q-th cohomology group of U with values in F** *by*

$$H^q(U, F) = R^q \Gamma_U(F).$$

Instead of $H^q(U, F)$ we sometimes write $H^q(T; U, F)$. If T has a final object e, the notation $H^q(T, F)$ instead of $H^q(T; e, F)$ is also common.

Example. Let X be a topological space, let T denote its topology (cp. (1.1)), and let F be an abelian sheaf on X, i.e. on T. Then X is a final object of T, and the cohomology groups $H^q(X, F)$, which we just defined, are the usual cohomology groups obtained from the so-called standard resolution of F (cp. [13], ch. II, 4.3, 4.4, 7.2.1).

(3.3.2) Example. Let G be a group, and let T_G be the canonical topology on the category of left G-sets (cp. (1.3.2)). We showed (1.3.2.2) that the category of left G-modules is equivalent to the category of abelian sheaves on T_G. This equivalence is provided by the mutually quasi-inverse functors $A \mapsto \mathrm{Hom}_G(\,\cdot\,, A)$ and $F \mapsto F(G)$.

Let e be a one-element set with its unique structure as left G-set. For the abelian sheaf $\mathrm{Hom}_G(\,\cdot\,, A)$ belonging to a left G-module A we obtain

$$\Gamma_e(\mathrm{Hom}_G(\,\cdot\,, A)) = \mathrm{Hom}_G(e, A) = A^G.$$

Here A^G denotes the group of G-invariant elements in A. We see that under the above equivalence of categories the section functor Γ_e gets identified with the functor $A \mapsto A^G$. Therefore we obtain for the q-th cohomology group of e with values in $\mathrm{Hom}_G(\,\cdot\,, A)$

$$H^q(e, \mathrm{Hom}_G(\,\cdot\,, A)) = H^q(G, A),$$

where the right-hand side is the q-th cohomology group of G with coefficients in the G-module A, as defined e.g. via the standard resolution (cp. [31], ch. VII, §§1 – 3).

Conversely, we can use group cohomology to understand the cohomology groups $H^q(U, \mathrm{Hom}_G(\,\cdot\,, A))$ for an arbitrary left G-set U in the following way:

The G-set U is the disjoint union of the orbits U_i of G in U, and for each orbit U_i we have a (non-canonical) G-isomorphism $U_i \cong G/H_i$ with some subgroup H_i of G. Within the framework of categories the disjoint union corresponds to the direct sum, and therefore we obtain

$$\begin{aligned} H^q(U, \mathrm{Hom}_G(\,\cdot\,, A)) &= \prod H^q(U_i, \mathrm{Hom}_G(\,\cdot\,, A)) \\ &\cong \prod H^q(G/H_i, \mathrm{Hom}_G(\,\cdot\,, A)). \end{aligned}$$

For a subgroup H of G the application of the section functor $\Gamma_{G/H}$ to $\mathrm{Hom}_G(\,\cdot\,, A)$ results in the following:

$$\Gamma_{G/H}(\mathrm{Hom}_G(\,\cdot\,, A)) = \mathrm{Hom}_G(G/H, A) = A^H.$$

Hence we obtain

$$H^q(G/H, \mathrm{Hom}_G(\,\cdot\,, A)) = H^q(H, A).$$

Here $H^q(H, A)$ denotes the q-th cohomology group of H with coefficients in A, which is naturally an H-module via the inclusion $H \subset G$.

(3.3.3) Example. Let G be a profinite group, and let T_G be the canonical topology on the category of continuous G-sets (cp. (1.3.3)). We proved in (1.3.3.2) that the category of continuous G-modules is equivalent to the category of abelian sheaves on T_G. This equivalence is provided by the mutually quasi-inverse functors $A \mapsto \mathrm{Hom}_G(\,\cdot\,, A)$ and $F \mapsto \varinjlim F(G/H)$, where H runs through all open normal subgroups of G.

Let us choose a one-element continuous G-set e. As in the previous example we see that under the above equivalence of categories the section functor Γ_e gets identified with the functor $A \mapsto A^G$. For the q-th cohomology group of e with values in the abelian sheaf $\mathrm{Hom}_G(\,\cdot\,, A)$ we obtain

$$H^q(e, \mathrm{Hom}_G(\,\cdot\,, A)) = H^q(G, A).$$

On the right-hand side we now have to take the q-th Tate cohomology group of the profinite group G with coefficients in the continuous G-module A, which can either be defined using **continuous** cochains of G with values in A, or as the inductive limit of the ordinary cohomology groups $H^q(G/H, A^H)$. (cp. [32], ch. I, § 2).

For an arbitrary continuous G-set U the cohomology groups $H^q(U, \mathrm{Hom}_G(\,\cdot\,, A))$ have a similar interpretation as in the previous example, the difference being that only cohomology groups of **open** subgroups of G occur.

(3.4) The Spectral Sequences for Čech Cohomology

The section functor $\Gamma_U : \mathcal{S} \to \mathcal{A}b$ factorizes as

$$\begin{aligned} \Gamma_U &= H^0(\{U_i \to U\}, \,\cdot\,) \circ i \\ &= \check{H}^0(U, \,\cdot\,) \circ i \end{aligned}$$

(cp. (2.2.2) and (2.2.8)). Here $i : \mathcal{S} \to \mathcal{P}$ is the inclusion, which is left exact (cp. (3.2.1) ii)). Let us now consider the right derived functors of i and define

(3.4.1) $$R^q i =: \mathcal{H}^q(\,\cdot\,).$$

For each $F \in \mathcal{S}$ we obtain an abelian presheaf $\mathcal{H}^q(F)$, in particular $\mathcal{H}^0(F) = F$, viewed as an abelian presheaf.

(3.4.2) Proposition. *For each abelian sheaf F and each $U \in T$ we have a canonical isomorphism*

$$\mathcal{H}^q(F)(U) \cong H^q(U, F).$$

Proof: The map $U \mapsto H^q(U,F)$ defines an abelian presheaf on T as follows: A morphism $U \to V$ in T induces a morphism $\Gamma_V \to \Gamma_U$ of functors, and therefore also a morphism $R^q\Gamma_V \to R^q\Gamma_U$ of ∂-functors (cp. (0.2.2)). Therefore we obtain homomorphisms $H^q(V,F) \to H^q(U,F)$, which are functorial in F.

To prove the proposition we show that the abelian presheaves $H^q(\,\cdot\,,F)$ are the derived functors of i.

We have $H^0(\,\cdot\,,F) = F = \mathcal{H}^0(F)$.

The functors $H^q(\,\cdot\,,F)$ form an exact ∂-functor from $\mathcal{S}$ to $\mathcal{P}$. To see this, let $0 \to F' \to F \to F'' \to 0$ be an exact sequence in $\mathcal{S}$. We then have the connecting homomorphisms $H^q(U,F'') \to H^{q+1}(U,F')$, which are functorial in U. Hence we obtain a morphism of abelian presheaves

$$H^q(\,\cdot\,,F'') \to H^{q+1}(\,\cdot\,,F'),$$

which is functorial in short exact sequences from $\mathcal{S}$, and the long cohomology sequence attached to a short exact sequence from $\mathcal{S}$ is exact in $\mathcal{P}$.

Moreover, for $q > 0$ and for each injective object F from $\mathcal{S}$ we have $H^q(\,\cdot\,,F) = 0$.

Therefore $F \mapsto (H^q(\,\cdot\,,F))_{q\geq 0}$ is a universal exact ∂-functor from $\mathcal{S}$ to $\mathcal{P}$, such that $H^0(\,\cdot\,,F) = i(F)$. Hence this functor coincides with $(R^q i)_{q\geq 0}$ (cp. (0.2.2)). □

(3.4.3) Proposition. *For each abelian sheaf F on T we have*

$$\mathcal{H}^q(F)^\dagger = 0 \quad \textit{for} \quad q > 0,$$

or, more explicitly (cp. (3.1)),

$$\check{H}^0(U, \mathcal{H}^q(F)) = 0 \quad \textit{for} \quad U \in T,\ q > 0.$$

Proof: From (3.1.3) i) and ii) we know that the canonical morphism $\mathcal{H}^q(F)^\dagger \to \mathcal{H}^q(F)^\#$ is a monomorphism. Hence it suffices to show that $\mathcal{H}^q(F)^\# = 0$ for $q > 0$.

Let us consider the following factorization of the identical functor $id_\mathcal{S}$ of $\mathcal{S}$:

$$\mathcal{S} \xrightarrow{i} \mathcal{P} \xrightarrow{\#} \mathcal{S}.$$

By (3.2.1) ii), the functor # is exact. Therefore $R^q\# = 0$ for $q > 0$, and in particular each object in $\mathcal{P}$ is #-acyclic. Hence there exists for each $F \in \mathcal{S}$ a spectral sequence

$$E_2^{pq} = R^p\#(\mathcal{H}^q(F)) \Longrightarrow E^{p+q} = R^{p+q}id_\mathcal{S}(F)$$

(cp. (0.2.3.5)), which is functorial in F. Since $R^p\# = 0$ for $p > 0$, we obtain $E_2^{pq} = 0$ for $p > 0$. Therefore the edge morphism (cp. (0.2.3.2))

$$E^q \to E_2^{0,q}$$

of this spectral sequence is an isomorphism for all q (cp. (0.2.3.4)). This means that

$$R^q id_\mathcal{S}(F) \cong \mathcal{H}^q(F)^\#$$

for all q. But for $q > 0$ we have $R^q id_\mathcal{S} = 0$, and therefore $\mathcal{H}^q(F)^\# = 0$ as well. □

(3.4.4) Theorem. *(The spectral sequence for Čech cohomology).*

i) Let $\{U_i \to U\}$ be a covering in T. For each $F \in \mathcal{S}$ there is a spectral sequence

$$E_2^{pq} = H^p(\{U_i \to U\}, \mathcal{H}^q(F)) \Longrightarrow E^{p+q} = H^{p+q}(U, F),$$

which is functorial in F.

ii) For each $F \in \mathcal{S}$ there is a spectral sequence

$$E_2^{pq} = \check{H}^p(U, \mathcal{H}^q(F)) \Longrightarrow E^{p+q} = H^{p+q}(U, F),$$

which is functorial in F.

Proof: We start with the following factorizations of the section functor Γ_U, mentioned already at the beginning of this section:

$$\mathcal{S} \xrightarrow{i} \mathcal{P} \xrightarrow{H^0(\{U_i \to U\}, \cdot)} Ab$$

$$\mathcal{S} \xrightarrow{i} \mathcal{P} \xrightarrow{\check{H}^0(U, \cdot)} Ab.$$

Since the left adjoint functor # of i is exact (cp. (3.3.1) ii)), i maps injective objects of $\mathcal{S}$ to injective objects of $\mathcal{P}$. This is proved in exactly the same way as statement ii) in (2.3.1). But injective objects in $\mathcal{P}$ are G-acyclic for each left exact additive functor $G : \mathcal{P} \to \mathcal{A}b$. The result now follows from Theorem (0.2.3.5). □

Let us have a closer look at the edge morphisms $E_2^{p,0} \to E^p$ belonging to the spectral sequences for Čech cohomology. These are the edge morphisms

$$(3.4.5) \qquad \begin{cases} H^p(\{U_i \to U\}, F) \to H^p(U,F) \\ \check{H}^p(U,F) \to H^p(U,F) \end{cases}$$

which are functorial in $F \in \mathcal{S}$. For an explicit description of these maps see [13], ch. II, 5.4.

(3.4.6) Corollary. *Let $\{U_i \to U\}_{i \in I}$ be a covering in T, and let F be an abelian sheaf such that $H^q(U_{i_0} \times_U \cdots \times_U U_{i_r}, F) = 0$ for all $q > 0$ and all $(i_0, \ldots, i_r) \in I^{r+1}$. Then the edge morphisms*

$$H^p(\{U_i \to U\}, F) \to H^p(U,F)$$

are isomorphisms for all p.

Proof: By (3.4.2) we have isomorphisms

$$\mathcal{H}^q(F)(U_{i_0} \times_U \cdots \times_U U_{i_r}) \cong H^q(U_{i_0} \times_U \cdots \times_U U_{i_r}, F).$$

Consider the group $C^r(\{U_i \to U\}, \mathcal{H}^q(F))$ of r-cochains with values in the presheaf $\mathcal{H}^q(F)$:

$$C^r(\{U_i \to U\}, \mathcal{H}^q(F)) = \coprod_{(i_0,\ldots,i_r) \in I^{r+1}} \mathcal{H}^q(F)\,(U_{i_0} \times_U \cdots \times_U U_{i_r}) = 0.$$

Since the Čech cohomology groups $H^p(\{U_i \to U\}, \mathcal{H}^q(F))$ coincide by (2.2.3) with the cohomology groups of the cochain complex $C^*(\{U_i \to U\}, \mathcal{H}^q(F))$, we obtain

$$H^p(\{U_i \to U\}, \mathcal{H}^q(F)) = 0$$

for $q > 0$. Hence in the spectral sequence

$$E_2^{pq} = H^p(\{U_i \to U\}, \mathcal{H}^q(F)) \Longrightarrow E^{p+q} = H^{p+q}(U,F)$$

all terms E_2^{pq} with $q > 0$ vanish. Therefore the edge morphism

$$E_2^{p,0} = H^p(\{U_i \to U\}, F) \to E^p = H^p(U, F)$$

is an isomorphism for all p (cp. (0.2.3.4)). □

For the edge morphisms $\check{H}^p(U, F) \to H^p(U, F)$ we have the following general result:

(3.4.7) Corollary. *For all abelian sheaves F the homomorphisms*

$$\check{H}^p(U, F) \to H^p(U, F)$$

are bijective for $p = 0, 1$ and injective for $p = 2$.

Proof: The case $p = 0$ is obvious. The terms $E_2^{0q} = \check{H}^0(U, \mathcal{H}^q(F))$ in the spectral sequence

$$E_2^{pq} = \check{H}^p(U, \mathcal{H}^q(F)) \Longrightarrow E^{p+q} = H^{p+q}(U, F)$$

vanish by (3.4.3) for all $q > 0$. Consider the exact sequence of terms of low degree

$$0 \to E_2^{1,0} \to E^1 \to E_2^{0,1} \to E_2^{2,0} \to E^2$$

(cp. (0.2.3.2)). The result follows, since $E_2^{0,1} = 0$. □

(3.4.8) Remark. Because of (0.2.3.3) and (3.4.3), Corollary (3.4.7) may be generalized as follows: If $\check{H}^p(U, \mathcal{H}^q(F)) = 0$ for $0 < q < n$, the map

$$\check{H}^m(U, F) \to H^m(U, F)$$

is an isomorphism for $m \leq n$ and a monomorphism for $m = n + 1$.

Exercise. Let G be a profinite group and let T_G be the canonical topology on the category of continuous G-sets. The category of abelian sheaves on T_G is equivalent to the category of continuous G-modules via the mutually quasi-inverse functors $F \to \varinjlim F(G/H)$ and $A \to \mathrm{Hom}_G(\,\cdot\,, A)$. Let e be a one-element G-set. Show the following:

i) Let H be an open normal subgroup of G. Then the above equivalence of categories transforms the spectral sequence

$$E_2^{pq} = H^p(\{G/H \to e\}, \mathcal{H}^q(F)) \Longrightarrow E^{p+q} = H^{p+q}(e, F)$$

(cp. (3.4.4)) into the spectral sequence

$$E_2^{pq} = H^p(G/H, H^q(H,A)) \Longrightarrow E^{p+q} = H^{p+q}(G,A),$$

known as the Hochschild-Serre spectral sequence for the normal subgroup H of G (cp. (3.7.9)).

ii) The spectral sequence

$$E_2^{pq} = \check{H}^p(e, \mathcal{H}^q(F)) \Longrightarrow E^{p+q} = H^{p+q}(e,F)$$

(cp. (3.4.4)) is trivial, which means $E_2^{pq} = 0$ for $q > 0$. Therefore the edge morphisms $E_2^{p,0} \to E_2^p$ are isomorphisms, i.e.

$$\check{H}^p(e,F) \cong H^p(e,F).$$

(3.5) Flabby Sheaves

(3.5.1) Definition. *An abelian sheaf F on T is called* **flabby** *(or* **flask***), if $H^q(\{U_i \to U\}, F) = 0$ for $q > 0$ and all coverings $\{U_i \to U\}$ in T.*

(3.5.2) Proposition. *i) Let $0 \to F' \to F \to F'' \to 0$ be an exact sequence in $\mathcal{S}$. If F' is flabby, the sequence is exact in $\mathcal{P}$ as well.*

ii) Let $0 \to F' \to F \to F'' \to 0$ be an exact sequence in $\mathcal{S}$. If F' and F are flabby, so is F''.

iii) If the direct sum $F \oplus G$ of abelian sheaves is flabby, so is F.

iv) Injective abelian sheaves are flabby.

Proof: i) For each $U \in T$ there is an exact sequence

$$0 \to F'(U) \to F(U) \to F''(U) \to H^1(U,F') \to \cdots .$$

Since F' is flabby, (3.4.7) implies that

$$H^1(U,F') \cong \check{H}^1(U,F') = \lim_{\{U_i \to U\}} H^1(\{U_i \to U\}, F') = 0.$$

Therefore the sequence $0 \to F' \to F \to F'' \to 0$ is exact in $\mathcal{P}$ as well (cp. (2.1.1)).

ii) Since by i) the sequence $0 \to F' \to F \to F'' \to 0$ is exact in $\mathcal{P}$ as well, we obtain for each covering $\{U_i \to U\}$ the exact cohomology sequence

$$\cdots \to H^q(\{U_i \to U\}, F) \to H^q(\{U_i \to U\}, F'') \\ \to H^{q+1}(\{U_i \to U\}, F') \to \cdots ,$$

from which the statement about F'' is obvious.

iii) The claim follows from

$$H^q(\{U_i \to U\}, F \oplus G) = H^q(\{U_i \to U\}, F) + H^q(\{U_i \to U\}, G).$$

iv) This is obvious from the definition of the Čech cohomology groups $H^q(\{U_i \to U\}, F)$. □

(3.5.3) Corollary. *For an abelian sheaf F on T the following are equivalent:*
i) F is flabby.

ii) For all $q > 0$ we have $\mathcal{H}^q(F) = 0$, and therefore $H^q(U, F) = 0$ (cp. (3.4.2)).

In particular, flabby resolutions in $\mathcal{S}$ can be used to compute $\mathcal{H}^q(\,\cdot\,)$ and $H^q(U, \cdot\,)$.

Proof: ii) $\Longrightarrow$ i): Let $\{U_i \to U\}_{i \in I}$ be a covering in T. For all $q > 0$ and all $(i_0, \ldots, i_r) \in I^{r+1}$ we have

$$H^q(U_{i_0} \times_U \cdots \times_U U_{i_r}, F) \cong \mathcal{H}^q(F)(U_{i_0} \times_U \cdots \times_U U_{i_r}) = 0.$$

Therefore, by (3.4.6), the edge morphisms

$$H^p(\{U_i \to U\}, F) \to H^p(U, F)$$

are isomorphisms for all p. In particular, for $p > 0$, $H^p(\{U_i \to U\}, F) = 0$, and hence F is flabby.

i) $\Longrightarrow$ ii): Let $0 \to F \to M^0 \to M^1 \to \cdots$ be an injective resolution of the flabby sheaf F in $\mathcal{S}$. We have to show that this sequence is also exact in $\mathcal{P}$ (see the construction of right derived functors in (0.2.2.1)).

Let $Z^i = ker(M^i \to M^{i+1})$ for $i \geq 0$. We obtain the exact sequences

$$0 \to Z^0 \to M^0 \to Z^1 \to 0$$
$$0 \to Z^1 \to M^1 \to Z^2 \to 0$$
$$\cdots$$
$$0 \to Z^i \to M^i \to Z^{i+1} \to 0$$
$$\cdots$$

in the category $\mathcal{S}$. Since $Z^0 \cong F$ is flabby, the sequence $0 \to Z^0 \to M^0 \to Z^1 \to 0$ is also exact in $\mathcal{P}$ by (3.5.2) i). Since M^0 is flabby as an injective sheaf by (3.5.2) iv), Z^1 is flabby as well by (3.5.2) ii), and we can continue. Therefore the sequence $0 \to F \to M^0 \to M^1 \to \cdots$ remains exact in the category $\mathcal{P}$.

The fact, that the vanishing of $\mathcal{H}^q(F)$ or $H^q(U, F)$ for $q > 0$ and flabby sheaves F implies, that we can use flabby resolutions for the computation of $\mathcal{H}^q(\,\cdot\,)$ or $H^q(U,\,\cdot\,)$, is proved as in [13], ch. II, 4.6, 4.7. □

(3.5.4) Example. On a given topology T all abelian sheaves will be flabby, if and only if the inclusion of categories $i : \mathcal{S} \to \mathcal{P}$ is exact. This follows from (3.5.3).

Here is an example: Let T be the canonical topology on the category of sets. By (1.3.2.2) the category of abelian sheaves on T gets identified with the category $\mathcal{A}b$ of abelian groups using the mutually quasi-inverse functors $F \to F(e)$ (e = one-element set) and $A \to \operatorname{Hom}(\,\cdot\,, A)$. Given an exact sequence

$$0 \to A' \to A \to A'' \to 0$$

of abelian groups, the formula $\operatorname{Hom}(U, A) = \prod_U A$ and the exactness of the functor $\prod$ imply, that the sequence remains exact in the category of abelian presheaves on T. Hence every abelian sheaf on T is flabby.

(3.6) The Functors f^s and f_s

Let $f : T \to T'$ be a morphism of topologies (cp. (1.2.2)). Let $\mathcal{S}$ and $\mathcal{P}$ denote the categories of abelian sheaves or presheaves on T respectively, and let $\mathcal{S}'$ and $\mathcal{P}'$ denote the corresponding categories with respect to T'.

We consider the additive functors

$$(3.6.1) \qquad \begin{cases} f^s : \mathcal{S}' \to \mathcal{S} \\ f_s : \mathcal{S} \to \mathcal{S}' \end{cases}$$

defined in an obvious notation as composites $f^s = \# \circ f^p \circ i'$ and $f_s = \#' \circ f_p \circ i$ (cp. (2.3), (3.1.1)). It is easily checked that for $F' \in \mathcal{S}'$ the image $f^p(i'(F'))$ is already a sheaf, so that $f^s = f^p \circ i'$.

(3.6.2) Proposition. *Let $f : T \to T'$ be a morphism of topologies. Then*

i) f_s is left adjoint to f^s.

ii) f^s is left exact.

iii) f_s is right exact and commutes with inductive limits. If f_s is actually exact, the functor f^s maps injective objects from $\mathcal{S}'$ to injective objects in $\mathcal{S}$.

Proof: i) follows, since $\#'$ is left adjoint to i' (cp. (3.1.1)) and f_p is left adjoint to f^p (cp. (2.3.1)).

ii) follows, since i' and f^p are left exact (cp. (3.2.1), ii) and (2.3)).

iii) Being left adjoint to f^s, the functor f_s is right exact and commutes with inductive limits (cp. (0.3.1.3)). The remaining part is proved in the same way as the corresponding part in (2.3.1), ii). □

(3.6.3) Remark. It is clear from the remarks (2.3.2) and (3.1.5), that the functors f^s and f_s can be defined more generally on the category of sheaves of sets (see [2], exp. III, 1.).

(3.6.4) Example. Let $\pi : X' \to X$ be a continuous map of topological spaces and $\pi^{-1} : T \to T'$ the morphism induced on their topologies.

Let F' be an abelian sheaf on X' (i.e. on T'). By definition $(\pi^{-1})^s(F')$ is equal to the abelian sheaf $U \mapsto F'(\pi^{-1}(U))$ on X (i.e. on T). In other words, $(\pi^{-1})^s(F')$ is equal to the so-called **direct** image $\pi_* F'$ of F' under π (cp. [13], ch. II, 1.13.).

Now let F be an abelian sheaf on X. Then $(\pi^{-1})_s(F)$ is equal to the sheaf associated to the following presheaf on X':

$$U' \mapsto (\pi^{-1})_p(F)(U') = \varinjlim_{U' \subset \pi^{-1}(U)} F(U)$$

(cp. the general construction of f_p during the proof of (2.3.1)). In other words, $(\pi_s^{-1}(F)$ is equal to the so-called **inverse** image $\pi^* F$ of F under π (cp. [13], ch. II, 1.12.).

(3.6.5) Example. Let G be a group and let T_G be the canonical topology on the category of left G-sets. The mutually quasi-inverse functors $F \to F(G)$ and $A \to \mathrm{Hom}_G(\,\cdot\,, A)$ identify the category $\mathcal{S}_G$ of abelian sheaves on T_G with the category of left G-modules (cp. (1.3.2.2)).

Let $\pi : G' \to G$ be a homomorphism of groups. Each left G-set becomes a left G'-set via $\pi : G' \to G$. Therefore we obtain a morphism $\tilde{\pi} : T_G \to T_{G'}$ of topologies, and hence functors

$$\begin{cases} \pi_* : \mathcal{S}_{G'} \to \mathcal{S}_G \\ \pi^* : \mathcal{S}_G \to \mathcal{S}_{G'} \end{cases}$$

between the categories $\mathcal{S}_G$ and $\mathcal{S}_{G'}$ of left G-modules or left G'-modules respectively. Here we use the notation $\pi_* = \tilde{\pi}^s$, $\pi^* = \tilde{\pi}_s$ analogous to (3.6.4). The functors π_* and π^* are given as follows:

For a left G'-module A' the left G-module $\pi_* A'$ equals $\mathrm{Hom}_{G'}(G, A')$, where G acts on $\pi_* A'$ by $(ga)(h) = a(hg)$, $g, h \in G$, $a \in \pi_* A'$.

If A is a left G-module, we have

$$\pi^* A = A$$

with the left G'-module structure induced by $\pi : G' \to G$. This follows from (2.3.3), which implies that the presheaf $\tilde{\pi}_p A$ on $T_{G'}$ is represented by the object $\tilde{\pi}(A) = A$, and hence in particular is already a sheaf.

In the special case that G' is a subgroup of G and that $\pi : G' \to G$ is the inclusion, the G-modules $\pi_* A'$ (for a given G'-module A') are well known in group cohomology (see e.g. [32], ch. I, 2.5., where $\pi_* A'$ is denoted by $M_G^{G'}(A')$).

If we specialize even further to $G' = \{1\}$, the G-modules $\pi_* A' = \mathrm{Hom}(G, A')$ are called **induced** G-**modules** (see [32], loc. cit.; in [31], ch. VIII, § 1 these G-modules are called co-induced).

(3.6.6) Example. If $\pi : G' \to G$ is a continuous homomorphism of profinite groups, we obtain, similar to the previous example (cp. (1.3.2.2)), functors

$$\begin{cases} \pi_* : \mathcal{S}_{G'} \to \mathcal{S}_G \\ \pi^* : \mathcal{S}_G \to \mathcal{S}_{G'} \end{cases}$$

between the categories $\mathcal{S}_G$ and $\mathcal{S}_{G'}$ of continuous left G-modules and G'-modules respectively.

If A' is a continuous G'-module, $\pi_* A'$ is the continuous G-module

$$\pi_* A' = \varinjlim \mathrm{Hom}_{G'}(G/H, A'),$$

where H runs through all open normal subgroups of G. Obviously, $\varinjlim \mathrm{Hom}_{G'}(G/H, A')$ is equal to the group of all continuous G'-maps $G \to A'$.

For a continuous G-module A we have

$$\pi^* A = A$$

with the continuous G'-module structure induced by $\pi : G' \to G$.

Again, in the special case $G' = \{1\}$, the G-modules $\pi_* A' = \varinjlim \mathrm{Hom}(G/H, A')$ for a given abelian group A' are called **induced G-modules** (see [32], loc. cit.).

Remark. For induced G-modules, where G is an arbitrary or a profinite group, the following is true: Each induced G-module is flabby in the canonical topology T_G. This follows from (3.5.4) and the fact that in general f^s maps flabby sheaves to flabby sheaves. We prove this in the next section (Proposition (3.7.2)).

In all three examples the functor f_s is in fact exact and not merely right exact. Here is a general result in this direction:

(3.6.7) Proposition. *Let T and T' be topologies, such that the underlying categories possess final objects and finite fibre products. Let*

$f : T \to T'$ be a morphism of topologies, which respects final objects and finite fibre products. Then $f_s : \mathcal{S} \to \mathcal{S}'$ is exact.

Proof: Since f_s is right exact (cp. (3.6.2), iii)), we have to show that f_s is left exact. Since $f_s = \#' \circ f_p \circ i$ and both i and $\#'$ are left exact (cp. (3.2.1), ii)), it suffices to prove left exactness for $f_p : \mathcal{P} \to \mathcal{P}'$ or for $F \to f_pF(U')$ for each $U' \in T'$ (cp. (2.1.1) ii)).

Recall the definition of $f_pF(U')$ from the proof of (2.3.1):

$$f_pF(U') = \varinjlim_{\mathcal{I}^0_{U'}} F_{U'}.$$

Here $\mathcal{I}_{U'}$ is the category of pairs (U, φ) with $U \in T$ and $\varphi : U' \to f(U)$ a morphism, and $F_{U'}$ denotes the covariant functor $(U, \varphi) \to F(U)$. Now the functor $F \to F_{U'}$ from $\mathcal{P}$ to the category $\mathcal{H}om(\mathcal{I}^0_{U'}, \mathcal{A}b)$ is exact (cp. (0.1.3.1), and therefore, in view of (0.2.3.1), it suffices to show that the dual category of $\mathcal{I}_{U'}$ is pseudofiltered. We verify that $\mathcal{I}_{U'}$ satisfies the properties dual to PS 1) and PS 2) (cp. (0.3.2)):

PS $1)^0$: Assume we are given a diagram

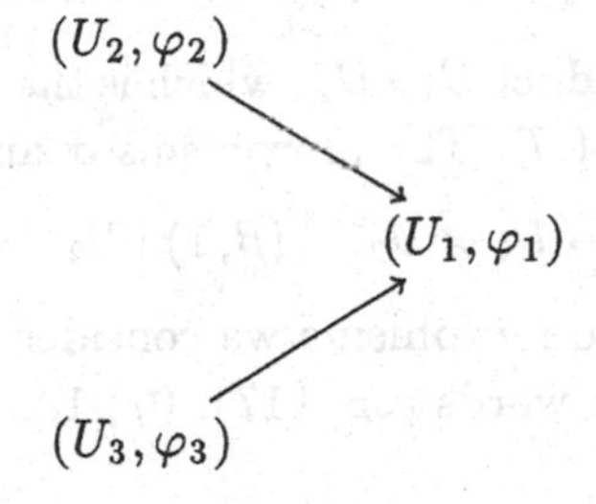

in the category $\mathcal{I}_{U'}$. By definition (cp. (2.3.1)) this is a diagram

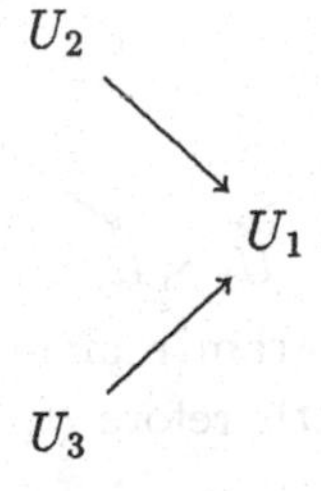

in T, so that the diagram

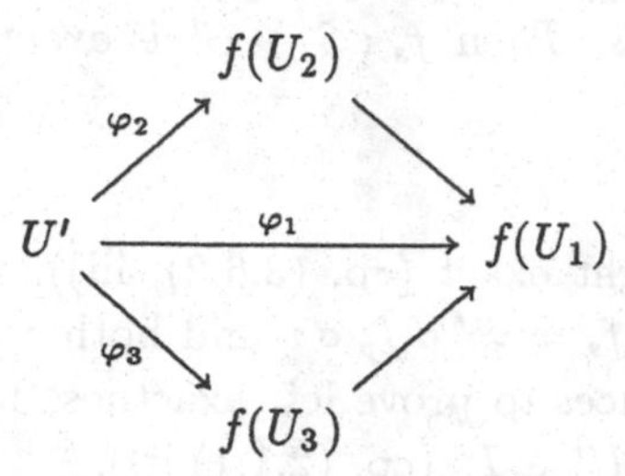

commutes. Consider the fibre product $U_2 \times_{U_1} U_3$ in T. By assumption on f we have $f(U_2 \times_{U_1} U_3) = f(U_2) \times_{f(U_1)} f(U_3)$. The commutative diagram from above then induces a morphism $\varphi' : U' \to f(U_2 \times_{U_1} U_3)$, and we obtain the following commutative diagram in $\mathcal{I}_{U'}$:

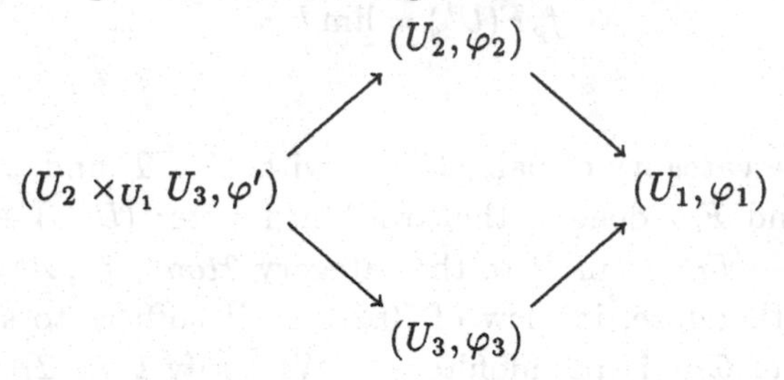

PS 2)^0: Assume we have a diagram

$$(U_2, \varphi_2) \rightrightarrows (U_1, \varphi_1)$$

in $\mathcal{I}_{U'}$. Thus we have morphisms $U_2 \overset{\alpha}{\underset{\beta}{\rightrightarrows}} U_1$ in T, such that $f(\alpha)\varphi_2 = \varphi_1 = f(\beta)\varphi_2$. We form the product $U_1 \times U_2$, which is the fibre product of U_1 and U_2 over the final object of T. The morphisms α and β induce morphisms

$$(\alpha, 1) : U_2 \to U_1 \times U_2, \quad (\beta, 1) : U_2 \to U_1 \times U_2,$$

and with respect to these morphisms we consider the fibre product $U = U_2 \times_{(U_1 \times U_2)} U_2$. In other words (cp. [17], 0_I, 1.2.4.) we have a cartesian diagram

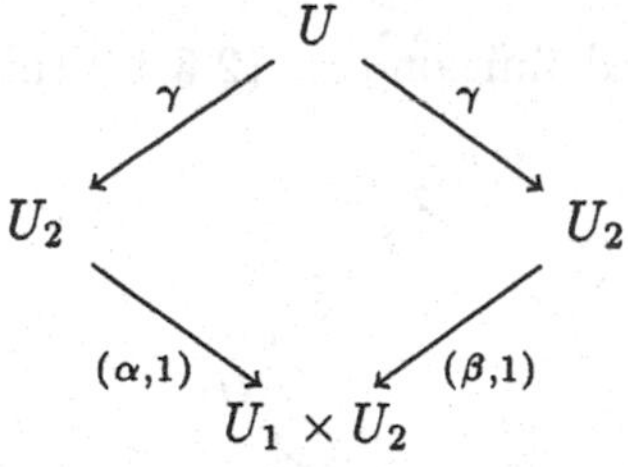

By assumption on f, this diagram remains cartesian after applying f. Since $f(\alpha)\varphi_2 = f(\beta)\varphi_2$, and therefore $f(\alpha, 1)\varphi_2 = f(\beta, 1)\varphi_2$, there is

a unique morphism $\varphi' : U' \to f(U)$, such that $f(\gamma)\varphi' = \varphi_2$. Then $(U, \varphi') \to (U_2, \varphi_2)$ does what we want, namely both composites

$$(U, \varphi') \to (U_2, \varphi_2) \rightrightarrows (U_1, \varphi_1)$$

coincide. □

Remark. The dual category of $\mathcal{I}_{U'}$ is in fact even filtered (cp. (0.3.2)): Given objects (U_1, φ_1) and (U_2, φ_2) in $\mathcal{I}_{U'}$, we can take $(U_1 \times U_2, (\varphi_1, \varphi_2))$.

(3.7) The Leray Spectral Sequences

Let $f : T \to T'$ be a morphism of topologies. From (3.6.2) ii) we know that the additive functor $f^s : \mathcal{S}' \to \mathcal{S}$ is left exact, and its right derived functors $R^q f^s : \mathcal{S}' \to \mathcal{S}$ exist.

(3.7.1) Proposition. *For each abelian sheaf $F' \in \mathcal{S}'$ there is an isomorphism*

$$R^q f^s(F') \cong (f^p \mathcal{H}^q(F'))^\#.$$

In other words, $R^q f^s(F')$ is the sheaf associated to the presheaf $U \mapsto H^q(f(U), F')$ on T (cp. (3.4.2), (2.3)).

Proof: The functor f^s splits into the left exact functor i' and the exact functor $\# \circ f^p$:

$$\mathcal{S}' \xrightarrow{i'} \mathcal{P}' \xrightarrow{\# \circ f^p} \mathcal{S}.$$

We have the following spectral sequence (cp. (0.2.3.5)):

$$E_2^{pq} = R^p(\# \circ f^p)(\mathcal{H}^q(F')) \Longrightarrow E^{p+q} = R^{p+q} f^s(F').$$

Now $E_2^{pq} = 0$ for $p > 0$, since $\# \circ f^p$ is exact. Hence the edge morphisms

$$E_2^{0,q} = (f^p \mathcal{H}^q(F'))^\# \leftarrow E^q = R^q f^s(F')$$

are isomorphisms for all q (cp. (0.2.3.4)). □

Since the left adjoint functor $f_s : \mathcal{S} \to \mathcal{S}'$ of f^s is not necessarily exact, we do not know in general, whether f^s maps injective objects in $\mathcal{S}'$ to injective objects in $\mathcal{S}$. However, the following is true:

(3.7.2) Proposition. *If $F' \in \mathcal{S}'$ is flabby, the same is true for $f^s F' \in \mathcal{S}$.*

Proof: We have to show that $H^q(\{U_i \to U\}, f^s F') = 0$ for all $q > 0$ and for all coverings $\{U_i \to U\}$ in T (cp. (3.5.1)). Using (2.2.3) this is immediate from the following identity for the Čech cochains:

$$\begin{aligned} C^n(\{U_i \to U\}, f^s F') &= \prod_{(i_0,\dots,i_n)} f^s F'(U_{i_0} \times_U \cdots \times_U U_{i_n}) \\ &= \prod_{(i_0,\dots,i_n)} F'(f(U_{i_0}) \times_{f(U)} \cdots \times_{f(U)} f(U_{i_n})) \\ &= C^n(\{f(U_i) \to f(U)\}, F'). \end{aligned}$$

□

(3.7.3) Corollary. *Let $T'' \xrightarrow{g} T \xrightarrow{f} T'$ be morphisms of topologies. For each flabby abelian sheaf F' on T' the sheaf $f^s F'$ on T is a g^s-acyclic object. This means: $R^q g^s(f^s F') = 0$ for $q > 0$.*

Proof: By (3.7.1) we have

$$R^q g^s(f^s F') = (g^p(\mathcal{H}^q(f^s F')))^{\#''}.$$

Since F' is flabby, $f^s F'$ is flabby as well by (3.7.2), hence by (3.5.3) we obtain $\mathcal{H}^q(f^s F') = 0$ for $q > 0$ and therefore indeed $R^q g^s(f^s F') = 0$ for $q > 0$. □

For the composition $T'' \xrightarrow{g} T \xrightarrow{f} T'$ of morphisms of topologies the following formulae are immediate from the definition or from (3.6.2) i):

$$(3.7.4) \qquad \begin{cases} (fg)^s = g^s f^s \\ (fg)_s = f_s g_s \end{cases}$$

Since by (3.7.3) and (3.5.2) iv) the sheaves $f^s F'$ are g^s-acyclic for an injective object $F' \in \mathcal{S}'$, the general existence theorem (0.2.3.5) for spectral sequences is applicable, and we obtain:

(3.7.5) Theorem *(The Leray spectral sequence). Let $T'' \xrightarrow{g} T \xrightarrow{f} T'$ be morphisms of topologies. For all abelian sheaves F' on T' there is a spectral sequence*

$$E_2^{pq} = R^p g^s(R^q f^s(F')) \Longrightarrow E^{p+q} = R^{p+q}(fg)^s(F'),$$

which is functorial in F'.

Let $f : T \to T'$ be a morphism and let $U \in T$. Let $g : P \to T$ be the unique morphism from the discrete topology P (cp. (2.3.4)) to T, which maps the single object in P to U. Then we have $g^s = \Gamma_U$ and $(fg)^s = \Gamma_{f(U)}$, hence

$$R^n g^s(F) = H^n(U,F) \qquad \text{for } F \in \mathcal{S}$$
$$R^n(fg)^s(F') = H^n(f(U),F') \quad \text{for } F' \in \mathcal{S}'$$

The result is the following special case of (3.7.5):

(3.7.6) Theorem *(The Leray spectral sequence). Let $f : T \to T'$ be a morphism of topologies and let $U \in T$. For all abelian sheaves F' on T' there is a spectral sequence*

$$E_2^{pq} = H^p(U, R^q f^s(F')) \Longrightarrow E^{p+q} = H^{p+q}(f(U),F'),$$

which is functorial in F'.

For each abelian sheaf F' on T' the Leray spectral sequence (3.7.6) induces edge morphisms (cp. (0.2.3.2)) for $p \geq 0$, which are functorial in F':

(3.7.7)
$$H^p(f(U),F') \to H^0(U, R^p f^s(F'))$$
$$H^p(U, f^s F') \to H^p(f(U),F').$$

The interesting edge morphisms are $H^p(U, f^s F') \to H^p(f(U),F')$, whereas the other edge morphisms can be interpreted as the canonical morphisms of the presheaves $f^p\mathcal{H}^p(F')$ to their associated sheaves $R^p f^s(F')$ (cp. (3.7.1)).

(3.7.8) Example. Let $\pi : X' \to X$ be a continuous map of topological spaces, let $\pi^{-1} : T \to T'$ be the induced morphism of their topologies and let $\pi_* = (\pi^{-1})^s$ (cp. (3.6.4)).

For an abelian sheaf F' on X' the sheaf $R^q\pi_*(F')$ is the sheaf associated to the presheaf $U \mapsto H^q(\pi^{-1}(U),F')$, and the Leray spectral sequence reads

$$E_2^{pq} = H^p(X, R^q\pi_*(F')) \Longrightarrow E^{p+q} = H^{p+q}(X',F')$$

(see [13], ch. II, 4.17).

(3.7.9) Example. Let $\pi : G' \to G$ be a homomorphism of groups. We keep the notations of (3.6.5) and use the interpretation of group cohomology given in (3.3.2). The Leray spectral sequence for the morphism

$\tilde{\pi} : T_G \to T_{G'}$ induced by π, and for a one-element G-set yields the following spectral sequence in group cohomology:

For all left G'-modules A' there is a spectral sequence

$$E_2^{pq} = H^p(G, R^q\pi_*(A')) \Longrightarrow E^{p+q} = H^{p+q}(G', A'),$$

which is functorial in A'. Here π_* is the functor $A' \mapsto \mathrm{Hom}_{G'}(G, A')$ from the category of left G'-modules to the category of left G-modules.

As a special case of the Leray spectral sequence we obtain the so-called **Hochschild-Serre spectral sequence** (cp. [32], ch. I., 2.5., [31], ch. VII, 6.): Let H be a normal subgroup of G and let $\pi : G \to G/H$ be the natural homomorphism. For each G-module A we have

$$\pi_* A = \mathrm{Hom}_G(G/H, A) = A^H,$$

hence

$$R^q\pi_*(A) = H^q(H, A).$$

The above spectral sequence then reads

$$E_2^{pq} = H^p(G/H, H^q(H, A)) \Longrightarrow E^{p+q} = H^{p+q}(G, A).$$

The edge morphisms

$$\begin{cases} H^p(G, A) \to H^p(H, A)^{G/H} \\ H^p(G/H, A^H) \to H^p(G, A) \end{cases}$$

belonging to this spectral sequence can be interpreted as **restriction** and **inflation** (cp. [31], ch. VII, 5.), and the exact sequence of terms of low degree (cp. (0.2.3.3)) is

$$\begin{aligned} 0 \to H^1(G/H, A^H) &\overset{inf}{\to} H^1(G, A) \overset{res}{\to} H^1(H, A)^{G/H} \\ &\overset{tr}{\to} H^2(G/H, A^H) \overset{inf}{\to} H^2(G, A). \end{aligned}$$

Here $tr : H^1(H, A)^{G/H} \to H^2(G/H, A^H)$ is the so-called **transgression**.

As another special case we consider the inclusion $\pi : H \to G$ of a subgroup H of G. It is easy to see that in this case π_* is exact, and hence $E_2^{pq} = 0$ for $q > 0$. This yields **Shapiro's Lemma**, namely that the edge morphisms $E_2^{p,0} \to E^p$ are isomorphisms (cp. [32], ch. I, 2.5.).

(3.7.10) Example. If $\pi : G' \to G$ is a continuous homomorphism of profinite groups, we obtain the Leray spectral sequence in the framework of cohomology of profinite groups:

$$E_2^{pq} = H^p(G, R^q\pi_*(A')) \Longrightarrow E^{p+q} = H^{p+q}(G', A').$$

Here π_* is the functor $A' \mapsto \varinjlim \mathrm{Hom}_{G'}(G/H, A')$ from the category of continuous G'-modules to the category of continuous G-modules. Again, as a special case, we obtain the Hochschild-Serre spectral sequence for closed normal subgroups of G.

(3.7.11) Example. For a profinite group G the general relations between Tate cohomology and ordinary cohomology of G are described by a Leray spectral sequence:

Let G_0 denote the underlying abstract group of the profinite group G. We have an obvious morphism $f : T_G \to T_{G_0}$ of topologies, and therefore for each left G_0-module A the Leray spectral sequence

$$E_2^{pq} = H^p(G, R^q f^s(A)) \Longrightarrow E^{p+q} = H^{p+q}(G_0, A).$$

Here $H^*(G, \cdot\,)$ denotes Tate cohomology and $H^*(G_0, \cdot\,)$ ordinary cohomology of G. The functor f^s from the category of left G_0-modules to the category of continuous left G-modules can be computed as follows:

$$\begin{aligned} f^s A &= \varinjlim \mathrm{Hom}_G(G/H, A) \\ &= \varinjlim A^H \\ &= \bigcup A^H , \end{aligned}$$

where H runs through all open normal subgroups of G.

(3.8) Localization

Let T be a topology and let Z be an object in T. We introduce a topology on the category T/Z of all Z-objects of T (cp. [17], 0_I, 1.1.11) by defining a family $\{U_i \to U\}$ of Z-morphisms to be a covering in T/Z, if $\{U_i \to U\}$ is a covering in T. The functor $i : T/Z \to T$, which assigns to each object $U \to Z$ of T/Z the object U, is a morphism of topologies.

(3.8.1) Lemma. *The functor i^s from the category of abelian sheaves on T to the category of abelian sheaves on T/Z is exact.*

Proof: We have to show that $R^q i^s = 0$ for $q > 0$. By (3.7.1)

$$R^q i^s(F) = (i^p(\mathcal{H}^q(F)))^\#$$

for an abelian sheaf F on T. Now $\mathcal{H}^q(F)^{\#} = 0$ for $q > 0$ (cp. (3.4.3)), hence it suffices to show that i^p and $\#$ commute for presheaves C on T. Now $i^p C(U \to Z) = C(U)$ and therefore (cp. (3.1.3)):

$$\begin{aligned}(i^p C)^{\dagger}(U \to Z) &= \varinjlim_{\{U_i \to U\} \in Cov(T/Z)} H^0(\{U_i \to U\}, i^p C) \\ &= \varinjlim_{\{U_i \to U\} \in Cov(T)} H^0(\{U_i \to U\}, C) \\ &= C^{\dagger}(U) = i^p(C^{\dagger})(U \to Z).\end{aligned}$$

□

(3.8.2) Corollary. *For all abelian sheaves F on T there are functorial isomorphisms*

$$H^p(T/Z; U \to Z, i^s F) \cong H^p(T; U, F),$$

in particular

$$H^p(T/Z; Z \xrightarrow{id} Z, i^s F) \cong H^p(T; Z, F).$$

Proof: We consider the Leray spectral sequence (3.7.6) attached to the morphism $i : T/Z \to T$:

$$E_2^{pq} = H^p(T/Z; U \to Z, R^q i^s(F)) \Longrightarrow E^{p+q} = H^{p+q}(T; U, F).$$

By the previous lemma $E_2^{pq} = 0$ for $q > 0$, and therefore all edge morphisms $E_2^{p0} \to E^p$ are isomorphisms (cp. (0.2.3.4)). □

Example. Let G be a group and let T_G be the canonical topology on the category of left G-sets. For a subgroup H of G the set G/H of left cosets gH of G mod H is an object in T_G. It can be shown that the functor

$$\begin{cases} T_G/(G/H) & \to T_H \\ U \xrightarrow{\varphi} G/H & \mapsto \varphi^{-1}(1 \cdot H) \end{cases}$$

is an equivalence of topologies. (Here, a functor $f : T \to T'$ is in general called an **equivalence of topologies**, if f is an equivalence of categories, and both f and any functor quasi-inverse to f are morphisms of topologies.) A similar result holds for profinite groups G and open subgroups H of G.

(3.9) The Comparison Lemma

(3.9.1) Theorem. *Let $i : T' \to T$ be a morphism of topologies with the following properties:*

i) The functor $i : T' \to T$ is fully faithful and therefore we may identify T' with a full subcategory of T via i.

ii) A covering $\{U_i \to U\}$ in T with U_i and U objects in T' is a covering in T'.

iii) Each object U in T has a covering $\{U_i \to U\}$ with objects $U_i \in T'$.

Then the functors $i^s : \mathcal{S} \to \mathcal{S}'$ and $i_s : \mathcal{S}' \to \mathcal{S}$ are quasi-inverse equivalences between the categories $\mathcal{S}$ and $\mathcal{S}'$ of abelian sheaves on T resp. T'.

Proof: We show that the adjoint morphisms (cp. (0.1.1))

$$\rho : id_{\mathcal{S}'} \to i^s \circ i_s$$
$$\sigma : i_s \circ i^s \to id_{\mathcal{S}}$$

are isomorphisms.

First of all we show that for all $G \in \mathcal{S}'$ the morphism $\rho_G : G \to i^s i_s G$ is an isomorphism. Let $U \in T'$. From (3.6.1) and section (2.3) we obtain a canonical factorization of the homomorphism

$$\rho_G(U) : G(U) \to i^s i_s G(U) = (i_p G)^{\#}(U)$$

into the maps

$$G(U) \overset{(1)}{\to} i_p G(U) \overset{(2)}{\to} (i_p G)^{\#}(U).$$

Here

$$i_p G(U) = \varinjlim_{(V,\varphi)} G(V)$$

is the inductive limit of the contravariant functor $(V,\varphi) \mapsto G(V)$ on the category $\mathcal{I}_U$ of pairs (V,φ) with $V \in T'$ and $\varphi : U \to V$ a morphism. The morphism

$$G(U) \overset{(1)}{\to} i_p G(U)$$

is the canonical homomorphism corresponding to the pair (U, id_U). But (U, id_U) is an initial object in $\mathcal{I}_U$, hence a final object in $\mathcal{I}_U^0$, and therefore $G(U) \overset{(1)}{\to} i_pG(U)$ is an isomorphism (cp. (0.3.2.2)).

The morphism $i_pG(U) \overset{(2)}{\to} (i_pG)^{\#}(U)$ is the canonical morphism mapping a presheaf to its associated sheaf. By assumption iii) we find for each covering $\{U_i \to U\}$ in T a refinement

$$\{U'_j \to U\} \to \{U_i \to U\}$$

with objects U'_j from T'. By assumption ii) $\{U'_j \to U\}$ is a covering in T'. Finally, since T' is a full subcategory of T, we obtain from (0.3.3.1):

$$\begin{aligned}(i_pG)^{\dagger}(U) = \check{H}^0(U, i_pG) &= \varinjlim_{\{U_i \to U\} \in Cov(T)} H^0(\{U_i \to U\}, i_pG) \\ &= \varinjlim_{\{U'_j \to U\} \in Cov(T')} H^0(\{U'_j \to U\}, i_pG).\end{aligned}$$

As we already observed the map $G(U) \to i_pG(U)$ is an isomorphism for $U \in T'$. Since G is an abelian sheaf on T', we conclude from the identity above that the canonical map $(i_pG)(U) \to (i_pG)^{\dagger}(U)$ is an isomorphism for $U \in T'$. The same is then true for the map $(i_pG)(U) \to (i_pG)^{\#}(U)$, and hence the adjoint morphism $\rho_G : G \to i^s i_s G$ is an isomorphism for $G \in \mathcal{S}'$.

Consider next the adjoint morphism $\sigma_F : i_s i^s F \to F$ for $F \in \mathcal{S}$. We have $i_s i^s F = (i_p i^s F)^{\#}$, and again by (3.6.1) and section (2.3) for each $U \in T'$ the following diagram commutes:

$$\begin{array}{ccc} (i_p i^s F)^{\#}(U) & \xrightarrow{\sigma_F(U)} & F(U) \\ \uparrow{\scriptstyle (2)} & \nearrow{\scriptstyle (3)} & \\ (i_p i^s F)(U) & & \end{array}$$

Here again, (2) is the canonical morphism mapping a presheaf to its associated sheaf. The map

$$(i_p i^s F)(U) = \varinjlim_{(V,\varphi)} i^s F(V) \overset{(3)}{\to} F(U)$$

is induced by the maps $i^s F(\varphi) : i^s F(V) \to i^s F(U) = F(U)$. Since $U \in T'$, we can apply the result of the first part of the proof to the sheaf $G = i^s F \in \mathcal{S}'$: This yields that (2) is an isomorphism, and also the map (3) as reciprocal of $F(U) = i^s F(U) \xrightarrow{\cong} (i_p i^s F)(U)$. Hence for $U \in T'$ the map $\sigma_F(U)$ is an isomorphism as well.

Now let U be an arbitrary object in T. We select a covering $\{U_i \to U\}$ of U with U_i from T' and also coverings $\{U_{ij}^k \to U_i \times_U U_j\}$ with U_{ij}^k from T'. For any sheaf $F_0 \in \mathcal{S}$ the homomorphisms

$$F_0(U_i \times_U U_j) \to \prod_k F_0(U_{ij}^k)$$

are injective. The isomorphism $i_s i^s F(U) \to F(U)$ now follows from the commutative diagram with exact rows

$$\begin{array}{ccccccc} 0 \longrightarrow & i_s i^s F(U) & \longrightarrow & \prod_i i_s i^s F(U_i) & \longrightarrow & \prod_{i,j,k} i_s i^s F(U_{ij}^k) \\ & \downarrow & & \downarrow \cong & & \downarrow \cong \\ 0 \longrightarrow & F(U) & \longrightarrow & \prod_i F(U_i) & \longrightarrow & \prod_{i,j,k} F(U_{ij}^k) \end{array}$$

This finishes the proof of Theorem (3.9.1). □

During the first part of the proof of (3.9.1), when we established that $\rho_G : G \to i^s i_s G$ is an isomorphism, we only used the following assumptions, which resulted from i) – iii):

1.) $i : T' \to T$ is fully faithful.

2.) For each $U \in T'$ and each covering $\{U_i \to U\}$ in T there is a refinement $\{U'_j \to U\}$ of $\{U_i \to U\}$ in T'.

Under these weaker conditions we can even prove the exactness of i^s, not, however, that $\sigma_F : i_s i^s F \to F$ is an isomorphism: In fact, by 2.) the functors i^p and $\sharp$ commute for presheaves on T, and we obtain as in the proof of (3.8.1), that $R^q i^s = 0$ for $q > 0$. Therefore we have:

(3.9.2) Theorem. *Let $i : T' \to T$ be a morphism of topologies satisfying the following two assumptions:*

i) The functor $i : T' \to T$ is fully faithful (and therefore we identify T' with a full subcategory of T).

ii) For each $U \in T'$ and each covering $\{U_i \to U\}$ in T there exists a refinement $\{U'_j \to U\}$ of $\{U_i \to U\}$ in T'.

Then the adjoint morphism $\rho_G : G \to i^s i_s G$ is an isomorphism for all $G \in \mathcal{S}'$. Moreover, $i^s : \mathcal{S} \to \mathcal{S}'$ is exact.

Remark. The bijectivity of the adjoint morphism ρ_G for all $G \in \mathcal{S}'$ is equivalent to the fact that $i_s : \mathcal{S}' \to \mathcal{S}$ is fully faithful. This is easily seen from the commutative diagram for $G', G \in \mathcal{S}'$:

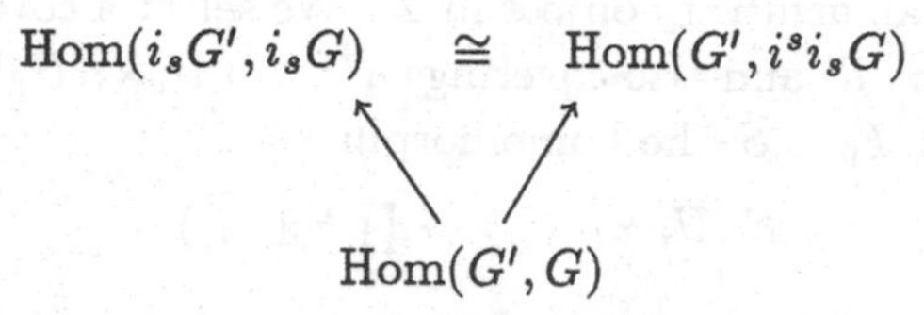

The horizontal arrow expresses the isomorphism given by the left adjointness of i_s to i^s (cp. (0.1.1)), and the right-hand arrow is induced by $\rho_G : G \to i^s i_s G$. To see that the diagram actually commutes, we observe the following: Let $v : G' \to G$ be given. Under the horizontal isomorphism the map $i_s(v) : i_s G' \to i_s G$ corresponds to a morphism $G' \to i^s i_s G$, which factors as $i^s i_s(v) \circ \rho_{G'} = \rho_G \circ v$, since ρ is functorial in G.

(3.9.3) Corollary. *Let $i : T' \to T$ be a morphism of topologies satisfying assumptions i) and ii) of (3.9.2), and let $U \in T'$. For all abelian sheaves F on T and all abelian sheaves F' on T', we have functorial isomorphisms*

$$H^p(T'; U, i^s F) \cong H^p(T; U, F)$$

and

$$H^p(T'; U, F') \cong H^p(T; U, i_s F').$$

Proof: The maps $H^p(T'; U, i^s F) \to H^p(T; U, F)$ are the edge morphisms $E_2^{p0} \to E^p$ of the Leray spectral sequence

$$E_2^{pq} = H^p(T'; U, R^q i^s F) \Longrightarrow E^{p+q} = H^{p+q}(T; U, F).$$

Since $R^q i^s = 0$ for $q > 0$, these edge morphisms are isomorphisms (cp. (0.2.3.4)).

The maps $H^p(T'; U, F') \to H^p(T; U, i_s F')$ for $F' \in \mathcal{S}'$ are the composites of the maps $H^p(T'; U, F') \to H^p(T'; U, i^s i_s F')$, induced by $\rho_{F'} : F' \to i^s i_s F'$, and the maps $H^p(T'; U, i^s i_s F') \to H^p(T; U, i_s F')$. The claim now follows from (3.9.2). □

(3.9.4) Example. Let G be a profinite group, and let T_G be the canonical topology on the category of continuous G-sets (cp. (1.3.3)). Let T'_G be the canonical topology on the category of **finite** continuous G-sets. The inclusion of categories yields a morphism $i : T'_G \to T_G$ of topologies, and we claim that it satisfies properties i) – iii) in (3.9.1): i) and ii) hold trivially. To check iii) we cover an arbitrary continous G-set U by the collection of all orbits Gu, $u \in U$, each of which is a **finite** continuous G-set, since the stabilizer of $u \in U$ is open in G.

From (1.3.3) and the comparison lemma (3.9.1) we obtain:

The category of abelian sheaves on T'_G is equivalent to the category of continuous G-modules under the mutually quasi-inverse functors $F \mapsto \varinjlim F(G/H)$ and $A \mapsto \mathrm{Hom}_G(\,\cdot\,, A)$.

This result remains true even if we consider on the category of finite continuous G-sets instead of the canonical topology T'_G the coarser topology T''_G. Here only **finite** families of morphisms $\{U_i \xrightarrow{\varphi_i} U\}$ with $U = \bigcup \varphi_i(U_i)$ are allowed as coverings. The reason is that T'_G is a so-called **noetherian** topology (cp. (3.10.1) and (3.10.3)).

Remark. In the case of an arbitrary group G we cannot argue in the same way, since property iii) in (3.9.1) need not hold.

(3.10) Noetherian Topologies

(3.10.1) Definition. *Let T be a topology. An object U of T is called* **quasicompact**, *if for each covering $\{U_i \to U\}_{i \in I}$ there exists a finite subset $I_0 \subset I$, such that the subfamily $\{U_i \to U\}_{i \in I_0}$ is still a covering in T. The topology T is called* **noetherian**, *if each object of T is quasicompact.*

If X is a topological space and if T is the topology of its open sets, the notion 'quasicompact' introduced above for $U \in T$ coincides with the standard notion for open sets. Moreover, using [3], ch. II, §4, no.2, prop. 9, we find that T is noetherian if and only if the topological space X is noetherian. Recall that this means that every non-empty closed subset of X has a minimal element with respect to inclusion.

If G is a profinite group, the topology T'_G (cp. (3.9.3)) is noetherian, but not the topology T_G. The quasicompact objects of T_G are precisely the finite continuous G-sets.

Let T be a topology. On the category underlying T we define another topology T^f allowing only the **finite** coverings $\{U_i \to U\}$ from T as coverings in T^f. The identical functor induces a morphism $i: T^f \to T$.

The functor $i^s : \mathcal{S} \to \mathcal{S}^f$ from the category $\mathcal{S}$ of abelian sheaves on T to the category $\mathcal{S}^f$ of abelian sheaves on T^f is given by $i^s F(U) = F(U)$ for $F \in \mathcal{S}$, $U \in T$ (cp. (3.6.1)). Obviously, i^s is fully faithful, but in general i^s is not an equivalence of categories.

(3.10.2) Proposition. *Let T be a noetherian topology. Then we have:*

i) $i^s : \mathcal{S} \to \mathcal{S}^f$ is an equivalence of categories.

ii) For all abelian sheaves on T there are ∂-functorial isomorphisms

$$H^q(T^f; U, i^s F) \cong H^q(T; U, F).$$

iii) An abelian sheaf F on T is flabby (cp. (3.5.1)) if and only if

$$H^q(\{U_i \to U\}, F) = 0$$

for all $q > 0$ and all **finite** *coverings $\{U_i \to U\}$ in T.*

Proof: i) We view $\mathcal{S}^f$ as a full subcategory of $\mathcal{S}$, and we have to show that any $F \in \mathcal{S}^f$ is also a sheaf on T. Now each covering $\{U_i \to U\}_{i \in I}$ in T has by assumption a finite subcovering $\{U_i \to U\}_{i \in I_0}$. For this the diagram

$$F(U) \to \prod_{i \in I_0} F(U_i) \rightrightarrows \prod_{i,j \in I_0} F(U_i \times_U U_j)$$

is exact. From (3.1.4) we conclude that F is a sheaf on T.

ii) This is an immediate consequence of i).

iii) By (3.5.3) we know that F is flabby if and only if $H^q(T; U, F) = 0$ for all $q > 0$ and all $U \in T$. Hence by ii) F is flabby on T if and only if F is flabby on T^f, which is the claim. □

As an immediate consequence of (3.10.2) i) we obtain the result mentioned at the end of example (3.9.4), since $T''_G = (T'_G)^f$.

(3.11) Commutation of the Functors $H^q(U, \cdot)$ with Pseudofiltered Inductive Limits

Let T be a topology, and let $\mathcal{S}$ be the category of abelian sheaves on T. If $F : \mathcal{I} \to \mathcal{S}$ $(i \mapsto F_i)$ is a functor from a category $\mathcal{I}$ to $\mathcal{S}$, we know by (3.2.3) i) that the inductive limit $\varinjlim F_i$ exists in $\mathcal{S}$ and equals the sheaf associated to the presheaf $U \mapsto \varinjlim F_i(U)$. The canonical morphisms $F_i \to \varinjlim F_i$ induce for each $U \in T$ and each $q \geq 0$ natural homomorphisms

$$\varinjlim H^q(U, F_i) \to H^q(U, \varinjlim F_i).$$

In general these homom[illegible] not isomorphisms. For $q = 0$ this is simply the fact that the presheaf li[illegible] need not be a sheaf. On the other hand we have:

(3.11.1) Theorem. *Let T be a noetherian topology and let $\mathcal{I}$ be a pseudofiltered category. Then the functors $H^q(U, \cdot)$ and $\varinjlim_{\mathcal{I}}$ commute.*

(3.11.2) Corollary. *If T is a noetherian topology, then the functor $H^q(U, \cdot)$ commutes with $\oplus$.*

Proof of (3.11.1): Let $i \mapsto F_i$ be the functor from the category $\mathcal{I}$ to the category $\mathcal{S}$ of abelian sheaves on T. Under the assumption in (3.11.1) on T and $\mathcal{I}$ we first show the following two facts:
i) The presheaf limit $U \mapsto \varinjlim F_i(U)$ is already a sheaf on T (i.e. (3.11.1) holds for $q = 0$).
ii) If each of the sheaves F_i is flabby (cp. (3.5.1)), then $\varinjlim F_i$ is flabby as well.

Let us denote the presheaf limit $U \mapsto \varinjlim F_i(U)$ by $p \cdot \varinjlim F_i$ for short. Let $\{U_j \to U\}$ be a **finite** covering in T. Since $\varinjlim$ commutes with direct sums, the group $C^n(\{U_j \to U\}, p \cdot \varinjlim F_i)$ of Čech n-cochains belonging to $\{U_j \to U\}$ can be computed as follows:

$$
\begin{aligned}
C^n(\{U_j \to U\}, p \cdot \varinjlim F_i) &= \bigoplus (p \cdot \varinjlim F_i)(U_{j_0} \times_U \cdots \times_U U_{j_n}) \\
&= \bigoplus \varinjlim F_i(U_{j_0} \times_U \cdots \times_U U_{j_n}) \\
&= \varinjlim (\bigoplus F_i(U_{j_0} \times_U \cdots \times_U U_{j_n}) \\
&= \varinjlim C^n(\{U_j \to U\}, F_i).
\end{aligned}
$$

Since F_i is an abelian sheaf, we have an exact sequence for each i:

$$0 \to F_i(U) \to C^0(\{U_j \to U\}, F_i) \xrightarrow{d^0} C^1(\{U_j \to U\}, F_i).$$

Since $\mathcal{I}$ is pseudofiltered, the application of $\varinjlim$ results again in an exact sequence (cp. (0.3.2.1)). Therefore we obtain the exact sequence

$$
\begin{aligned}
0 \to p \cdot \varinjlim F_i(U) &\to C^0(\{U_j \to U\}, p \cdot \varinjlim F_i) \\
&\to C^1(\{U_j \to U\}, p \cdot \varinjlim F_i).
\end{aligned}
$$

This proves that $p \cdot \varinjlim F_i$ is a sheaf on T^f, hence on T (cp. (3.10.2)).

If moreover F_i is flabby, the long sequence

$$0 \to F_i(U) \to C^0(\{U_j \to U\}, F_i) \xrightarrow{d^0} C^1(\{U_j \to U\}, F_i) \xrightarrow{d^1} \cdots$$

is exact (cp. (2.2.3)). Hence the same is true for the sequence after applying $\varinjlim$, which proves that $\varinjlim F_i$ is flabby (cp. (3.10.2) iii)). Thus statements i) and ii) are proven.

Assume for the moment that the functor $F : \mathcal{I} \to \mathcal{S}$ has a resolution

$$0 \to F \to F^0 \to F^1 \to \cdots$$

in the category $\mathcal{H}om(\mathcal{I}, \mathcal{S})$, such that all sheaves F_i^0, $F_i^1, \ldots$ are flabby in T. From ii) and (0.3.2.1) it follows that

$$0 \to \varinjlim F_i \to \varinjlim F_i^0 \to \varinjlim F_i^1 \to \cdots$$

is a flabby resolution of $\varinjlim F_i$ in $\mathcal{S}$. By (3.5.3) we can use this resolution to compute $H^q(U, \varinjlim F_i)$. For the section functor Γ_U we obtain by i):

$$\Gamma_U(\varinjlim F_i) = (\varinjlim F_i)(U) = \varinjlim F_i(U)$$

and similarly for the sheaves $\varinjlim F_i^0, \ldots$. Hence in fact

$$H^q(U, \varinjlim F_i) \cong \varinjlim H^q(U, F_i),$$

as was to be shown. □

The category $\mathcal{H}om(\mathcal{I}, \mathcal{S})$ is an abelian category with property Ab 5) and with generators (cp. (0.1.4.3)). Therefore (cp. (0.1.4.3)) F has an injective resolution

$$0 \to F \to F^0 \to F^1 \to \cdots$$

in $\mathcal{H}om(\mathcal{I}, \mathcal{S})$, and we claim that all $F_i^0, F_i^1, \ldots$ are injective (hence flabby). The easiest way to see this is the following:

If $\varphi : \mathcal{J} \to \mathcal{I}$ is a functor, we obtain by (2.3) functors

$$\mathcal{H}om(\mathcal{I}, \mathcal{P}) \underset{\varphi_p}{\overset{\varphi^p}{\rightleftarrows}} \mathcal{H}om(\mathcal{J}, \mathcal{P}).$$

Here $\varphi^p(E) = E \circ \varphi$, and φ_p is the left adjoint functor of φ^p. Let us take for φ the inclusion $\varphi : \{i\} \to \mathcal{I}$ from the discrete subcategory consisting only of the object i. Then we can compute $\varphi_p(E)$ for $E \in \mathcal{H}om(\{i\}, \mathcal{P}) \cong \mathcal{P}$ using (2.3.4) and obtain

$$\varphi_p(E)_j = \bigoplus_{\mathrm{Hom}_{\mathcal{I}}(i,j)} E.$$

As in (3.6.1) we also have the functors

$$\mathcal{H}om(\mathcal{I}, \mathcal{S}) \underset{\varphi_s}{\overset{\varphi^s}{\rightleftarrows}} \mathcal{H}om(\mathcal{J}, \mathcal{S})$$

induced by φ^p and φ_p via $i : \mathcal{S} \to \mathcal{P}$ and $\# : \mathcal{P} \to \mathcal{S}$. φ_s is left adjoint to φ^s. Again, for the inclusion $\varphi : \{i\} \to \mathcal{I}$ we can compute $\varphi_s(E)$ for $E \in \mathcal{H}om(\{i\}, \mathcal{S})$ and obtain

$$\varphi_s(E)_j = \Big(\bigoplus_{\mathrm{Hom}_{\mathcal{I}}(i,j)} E \Big)^{\#} = \bigoplus_{\mathrm{Hom}_{\mathcal{I}}(i,j)} E,$$

where the last equation is true by i). In particular, $\varphi_s : \mathcal{S} \to \mathcal{H}om(\mathcal{I}, \mathcal{S})$ is exact. Hence $\varphi^s : \mathcal{H}om(\mathcal{I}, \mathcal{S}) \to \mathcal{S}$ maps injective objects to injective objects, and we have shown our claim, since $\varphi^s(F) = F_i$.

Chapter II
Étale Cohomology

§ 1. The Étale Site of a Scheme

(1.1) Étale Morphisms

A morphism $f : X \to Y$ of schemes is called **locally finitely presented**, if for each point $x \in X$ there are affine open neighbourhoods V of $y = f(x)$ and U of x with $f(U) \subset V$, such that the $\Gamma(V, \mathcal{O}_Y)$-algebra $\Gamma(U, \mathcal{O}_X)$ is finitely presented (cp. [17], 6.2.).

The morphism $f : X \to Y$ is **flat** if for each point $x \in X$ the local ring $\mathcal{O}_{X,x}$ is a flat $\mathcal{O}_{Y,f(x)}$-module (cp. [20], 2.1.).

(1.1.1) A locally finitely presented morphism $f : X \to Y$ of schemes is called **étale** if it satisfies the following equivalent conditions:

i) f is flat and unramified. This means that for each point $y \in Y$ the fibre $X_y = X \times_Y \operatorname{spec}(k(y))$ as a $k(y)$-scheme is equal to the sum of spectra of finite separable field extensions of $k(y)$.

ii) For each affine scheme Y' over Y and each closed subscheme Y'_0 of Y', which is defined by a nilpotent ideal, the canonical map

$$\operatorname{Hom}_Y(Y', X) \to \operatorname{Hom}_Y(Y'_0, X)$$

is bijective (cp. [20], 17.1.1., 17.3.1., 17.6.2.).

(1.1.2) Proposition. *i) Open immersions are étale.*

ii) If $f : X \to Y$ and $g : Y \to Z$ are étale, so is the composite $g \circ f$.

iii) If $f : X \to X'$ and $g : Y \to Y'$ are étale S-morphisms, so is $f \times_S g : X \times_S Y \to X' \times_S Y'$.

iv) Let $f : X \to Y$ and $g : Y \to Z$ be morphisms. If $g \circ f$ and g are étale, so is f.

(cp. [20], 17.3.3.)

(1.2) The Étale Site

Let X be a scheme. We consider the category Et/X of étale X-schemes. By (1.1.2) the category Et/X has finite fibre products, and X is a final object in Et/X.

A family $\{X'_i \xrightarrow{\varphi_i} X'\}$ of morphisms in Et/X is called **surjective** provided $X' = \bigcup_i \varphi_i(X'_i)$. It is easy to check that the set of all surjective families of morphisms in Et/X satisfies the three axioms T1) – T3) of a topology (cp. I, 1.2.1.). We define the **étale site $X_{\text{ét}}$ of X** by

$$\begin{cases} cat(X_{\text{ét}}) & = Et/X \\ cov(X_{\text{ét}}) & = \text{set of surjective families} \\ & \quad\ \text{of morphisms in } Et/X. \end{cases}$$

The category of abelian sheaves on $X_{\text{ét}}$ is denoted by $\widetilde{X}_{\text{ét}}$. Sheaves on $X_{\text{ét}}$ are also called **étale sheaves** on X.

Remark. In [2], exp. VII the notation $\widetilde{X}_{\text{ét}}$ is used to denote the category of sheaves of sets on $X_{\text{ét}}$, the so-called **étale topos of X**, whereas the category of abelian sheaves is denoted by $(\widetilde{X}_{\text{ét}})_{ab}$.

For each abelian sheaf F on $X_{\text{ét}}$ and for each étale X-scheme X' the cohomology groups

$$H^q(X', F)$$

of X' with values in F are defined (cp. I, 3.3.1.). These cohomology groups are sometimes also denoted by $H^q_{\text{ét}}(X', F)$ or $H^q(X_{\text{ét}}; X', F)$.

(1.3) The Relation between Étale and Zariski Cohomology

Let X_{Zar} denote the topology of open sets of the scheme X. The inclusion

$$\varepsilon : X_{Zar} \to X_{\text{ét}}$$

is a morphism of topologies. For each abelian sheaf on $X_{\text{ét}}$ we have the Leray spectral sequence (I, 3.7.6.)

$$(1.3.1) \qquad E_2^{pq} = H^p_{Zar}(X, R^q\varepsilon^s(F)) \Longrightarrow E^{p+q} = H^{p+q}_{\text{ét}}(X, F),$$

which is functorial in F. Here $\varepsilon^s(F)$ denotes the abelian sheaf $U \mapsto F(U)$ on X_{Zar} (I, 3.6.4.).

This spectral sequence describes the relation between the étale cohomology and the Zariski cohomology on X. In general $R^q\varepsilon^s \neq 0$, so that the spectral sequence is non-trivial. This is already the case if $X = spec(K)$, K a field (cp. section 2).

(1.4) The Functors f_* and f^*

Let $f : X \to Y$ be a morphism of schemes. Then f induces a covariant functor from the category of étale Y-schemes to the category of étale X-schemes by

$$Y' \mapsto Y' \times_Y X.$$

This functor respects fibre products and surjective families of morphisms. In other words, f induces a morphism

$$f_{\text{ét}} : Y_{\text{ét}} \to X_{\text{ét}}$$

of topologies. The general theory (I, (3.6.1)) shows that the functors

$$\begin{aligned} f_* &= (f_{\text{ét}})^s : \widetilde{X}_{\text{ét}} \to \widetilde{Y}_{\text{ét}} \\ f^* &= (f_{\text{ét}})_s : \widetilde{Y}_{\text{ét}} \to \widetilde{X}_{\text{ét}} \end{aligned}$$

are defined for the categories of abelian sheaves on $X_{\text{ét}}$ resp. $Y_{\text{ét}}$.

For an abelian sheaf F on $X_{\text{ét}}$ the sheaf f_*F is called the **direct image of** F under $f : X \to Y$. By definition (I, (3.6.1)) we have for an étale Y-scheme Y':

$$f_*F(Y') = F(Y' \times_Y X).$$

For an abelian sheaf G on $Y_{\text{ét}}$ the sheaf f^*G on $X_{\text{ét}}$ is called the **inverse image of** G under $f : X \to Y$. By definition (I, (3.6.1)) f^*G is the sheaf associated to the presheaf $f^{\cdot}G = (f_{\text{ét}})_pG$. By I, (2.3.1) we have

$$f^{\cdot}G(X') = \varinjlim_{(\mathcal{I}_{X'})^0} G(Y')$$

for $X' \in X_{\text{ét}}$. Here $\mathcal{I}_{X'}$ is the category of pairs (Y', φ) consisting of an étale Y-scheme Y' and an X-morphism $\varphi : X' \to Y' \times_Y X$. Since $\mathrm{Hom}_X(X', Y' \times_Y X) = \mathrm{Hom}_Y(X', Y')$, we can also interpret $\mathcal{I}_{X'}$ as the category of Y-morphisms $X' \to Y'$ of X' to étale Y-schemes Y'. The dual category $(\mathcal{I}_{X'})^0$ is filtered, since $f_{\text{ét}}$ preserves fibre products and final objects (cp. the remark following the proof of I, (3.6.7)).

The general results of Chapter I (3.6.2, 3.6.7, 3.7.4) provide us with the following facts about f_* and f^*:

(1.4.2) i) f^* is left adjoint to f_*.

ii) f_* is left exact.

iii) f^* is exact and commutes with inductive limits.

iv) For morphisms $f : X \to Y$, $g : Y \to Z$ we have $(g \circ f)_* = g_* \circ f_*$, $(g \circ f)^* = f^* \circ g^*$.

If we apply the q-th right derived functor $R^q f_*$ of f_* to an abelian sheaf F on $X_{ét}$, we obtain (cp. I, (3.7.1)) the sheaf associated to the presheaf

$$Y' \mapsto H^q(Y' \times_Y X, F)$$

on $Y_{ét}$. Moreover:

(1.4.3) For each abelian sheaf F on $X_{ét}$ and each étale Y-scheme Y' we have the Leray spectral sequence

$$E_2^{pq} = H^p(Y', R^q f_*(F)) \Longrightarrow E^{p+q} = H^{p+q}(Y' \times_Y X, F),$$

which is functorial in F (cp. I, (3.7.6)).

From this spectral sequence we obtain for abelian sheaves $F \in \widetilde{X}_{ét}$ and $G \in \widetilde{Y}_{ét}$ canonical homomorphisms

$$(1.4.4) \qquad \begin{cases} H^p(Y', f_*F) & \to H^p(Y' \times_Y X, F) \\ H^p(Y', G) & \to H^p(Y' \times_Y X, f^*G), \end{cases}$$

which are functorial in F and G respectively. The first map is the edge morphism $E_2^{p0} \to E^p$, and the second is the composite of the homomorphism $H^p(Y', G) \to H^p(Y', f_*f^*G)$, induced from the adjoint morphism $\rho_G : G \to f_*f^*G$, with the edge morphism $H^p(Y', f_*f^*G) \to H^p(Y' \times_Y X, f^*G)$ for the sheaf f^*G.

(1.4.5) If $f : X \to Y$ and $g : Y \to Z$ are morphisms of schemes, we have for each abelian sheaf F on $X_{ét}$ the Leray spectral sequence

$$E_2^{pq} = R^p g_*(R^q f_*(F)) \Longrightarrow E^{p+q} = R^{p+q}(gf)_*(F),$$

which is functorial in F (cp. I, (3.7.5)).

As an application we will construct the so-called base change morphisms for cartesian diagrams using the edge morphisms

$$(1.4.6)\qquad \begin{cases} R^p g_*(f_*F) & \to R^p(gf)_*(F) \\ R^p(gf)_*(F) & \to g_*(R^p f_*(F)) \end{cases}$$

of the Leray spectral sequence:

Assume we are given a cartesian square

$$\begin{array}{ccc} X' & \xrightarrow{f'} & Y' \\ {\scriptstyle v'}\downarrow & & \downarrow{\scriptstyle v} \\ X & \xrightarrow{f} & Y \end{array}$$

of morphisms of schemes. Let F be an abelian sheaf on $X_{\text{ét}}$. Starting from the adjoint morphism $F \to v'_* v'^* F$ we consider the following composition of morphisms

$$\begin{aligned} R^p f_*(F) &\to R^p f_*(v'_* v'^* F) \\ &\overset{(1)}{\to} R^p(f \circ v')_*(v'^* F) = R^p(v \circ f')_*(v'^* F) \\ &\overset{(2)}{\to} v_*(R^p f_*(v'^* F)). \end{aligned}$$

Here (1) denotes the first and (2) the second edge morphism in (1.4.5). We obtain a morphism $R^p f_*(F) \to v_*(R^p f'_*(v'^* F))$, whose adjoint morphism (cp. (1.4.2) i))

$$(1.4.7)\qquad v^*(R^p f_*(F)) \to R^p f'_*(v'^* F))$$

is the so-called **base change morphism.** Obviously, it is functorial in F. If this base change morphism is an isomorphism for all cartesian diagrams over $f : X \to Y$ (i.e. for all base changes $v : Y' \to Y$), we say that $R^p f_*(F)$ commutes with arbitrary base change.

Let X' be an étale X-scheme, hence an object of $X_{\text{ét}}$. Let $f : X' \to X$ be the structure morphism. We define a morphism

$$X'_{\text{ét}} \to X_{\text{ét}}$$

of topologies by assigning to an étale X'-scheme Z' the étale X-scheme $Z' \to X' \xrightarrow{f} X$. Obviously, this morphism identifies $X'_{\text{ét}}$ with the localization $(X_{\text{ét}})/X'$ of $X_{\text{ét}}$ in X' (cp. I, (3.8)). For an abelian sheaf F on $X_{\text{ét}}$ we define $F/X' = f^*F$ and we claim that

$$(1.4.8)\qquad F/X'(Z') = F(Z')$$

for étale X'-schemes Z'.

By definition $F/X' = f^*F$ is the sheaf associated to the presheaf $f^{\cdot}F$ and by (1.4.1)

$$f^{\cdot}F(Z') = \varinjlim F(Z),$$

where the inductive limit extends over the dual category of $\mathcal{I}_{Z'}$. The category $\mathcal{I}_{Z'}$ can be viewed as the category of all X-morphisms $Z' \to Z$ of Z' to étale X-schemes Z and has the identical morphism $id_{Z'} : Z' \to Z'$ as initial object. Therefore (cp. 0, (3.3.2)) we have $f^{\cdot}F(Z') = F(Z')$ and $f^{\cdot}F$ is already a sheaf on $X'_{\text{ét}}$.

The sheaf F/X' is called **the restriction of F to $X'_{\text{ét}}$.** If we identify $X'_{\text{ét}}$ with the localization $(X_{\text{ét}})/X'$, we obtain from (1.4.8) and I, (3.8.2) canonical isomorphisms

(1.4.9) $$H^q(X_{\text{ét}}; X', F) \cong H^q(X'_{\text{ét}}; X', F/X').$$

(1.5) The Restricted Étale Site

A morphism $f : X \to Y$ of schemes is called **quasi-compact**, if for each quasi-compact open set $U \subset Y$ the inverse image $f^{-1}(U)$ is quasi-compact. $f : X \to Y$ is called **quasi-separated**, if the diagonal morphism $\Delta : X \to X \times_Y X$ is quasi-compact. A scheme X is called **quasi-separated**, if the morphism $X \to spec(\mathbb{Z})$ is quasi-separated.

A morphism $f : X \to Y$ is called **finitely presented**, if it is locally finitely presented, quasi-compact and quasi-separated (cp. [17], 6.3).

Let X be a scheme. We consider the category of **finitely presented** étale X-schemes. This category has fibre products and X is a final object. Again, we define a topology on this category by considering the surjective families of morphisms as coverings. This topology is called the **restricted étale site of X**, and we denote it by $X_{\text{ét.f.p.}}$.

(1.5.1) Lemma. *If X is quasi-compact, then the restricted étale site $X_{\text{ét.f.p.}}$ is noetherian (cp. I, (3.10.1)).*

Proof: We have to show that for a given covering $\{X'_i \xrightarrow{\varphi_i} X'\}_{i \in I}$ in $X_{\text{ét.f.p.}}$ there exists a finite subset $I_0 \subset I$, such that $\{X'_i \xrightarrow{\varphi_i} X'\}_{i \in I_0}$ is also a covering in $X_{\text{ét.f.p.}}$. By definition we have $X' = \bigcup \varphi_i(X'_i)$.

Each $\varphi_i : X'_i \to X'$ is open, since it is étale ([20], 2.4.6.), so that $X' = \bigcup_{i \in I} \varphi_i(X'_i)$ is an open covering of X'. Now $X' \to X$ is a finitely presented morphism, hence in particular quasi-compact. Therefore, since X is quasi-compact, X' is quasi-compact as well, and hence there exists a finite subset $I_0 \subset I$, such that $X' = \bigcup_{i \in I_0} \varphi_i(X'_i)$. □

The obvious inclusion of categories

$$i : X_{\text{ét.f.p.}} \to X_{\text{ét}}$$

is in fact a morphism of topologies. The general theory in I, (3.6.1) implies that we obtain functors

$$\begin{cases} res = i^s : \widetilde{X}_{\text{ét}} \to \widetilde{X}_{\text{ét.f.p.}} \\ ext = i_s : \widetilde{X}_{\text{ét.f.p.}} \to \widetilde{X}_{\text{ét}} \end{cases}$$

on the corresponding categories of abelian sheaves. The functor *res* (restriction) is left exact, the functor *ext* (extension) is exact and left adjoint to *res* (cp. I, (3.6.1), (3.6.7)).

(1.5.2) Proposition. *If the scheme X is quasi-separated, then the functors res : $\widetilde{X}_{\text{ét}} \to \widetilde{X}_{\text{ét.f.p.}}$ and ext : $\widetilde{X}_{\text{ét.f.p.}} \to \widetilde{X}_{\text{ét}}$ are mutually quasi-inverse equivalences of categories.*

Proof: We want to apply the general comparison theorem I, (3.9.1). Assumptions i) and ii), needed there, hold trivially, so that we concentrate on assumption iii). This reads in the current situation: Given an étale X-scheme X' we can find a covering $\{X'_i \to X'\}$ of X' by finitely presented étale X-schemes X'_i.

Let $f : X' \to X$ be the structure morphism, and let x' be a point of X'. Since $f : X' \to X$ is locally finitely presented, there are an open affine neighbourhood U of $x = f(x')$ in X and an open affine neighbourhood U' of x' in X', such that $f(U') \subset U$ and $\Gamma(U', \mathcal{O}_{X'})$ is a finitely presented $\Gamma(U, \mathcal{O}_X)$-algebra. The restriction

$$U' \to X$$

of $f : X' \to X$ to U' is étale, and we claim it is also finitely presented. This would imply the existence of a covering of the desired kind.

By construction the map $U' \to X$ splits into the finitely presented morphism $U' \to U$ and the canonical injection $U \to X$. Therefore we have to show that $U \to X$ is finitely presented. As an open immersion it is locally finitely presented and quasi-separated. Since X is quasi-separated by assumption and U is quasi-separated as an affine scheme, the morphism $U \to X$ is also quasi-compact ([17], 6.1.10.). □

From (1.5.1) and (1.5.2) and the theorem I, (3.11.1) we obtain:

(1.5.3) Corollary. *If X is a quasi-compact and quasi-separated scheme, then the functors $H^q_{ét}(X, \cdot)$ and $\varinjlim_I$ commute for pseudofiltered categories I. In particular, the functors $H^q_{ét}(X, \cdot)$ commute with direct sums.*

§ 2. The Case $X = spec(k)$

Let k be a field. Let $\bar{k}$ be a separable closure of k and let G denote the Galois group of $\bar{k}/k$ equipped with the canonical structure of a profinite group.

For each k-scheme X' we denote by $X'(\bar{k})$ the set of $\bar{k}$-valued points on X'/k, i.e. the set of k-morphisms $spec(\bar{k}) \to X'$. A $\bar{k}$-valued point of X' corresponds uniquely to a point $x' \in X'$ together with a k-homomorphism $k(x') \to \bar{k}$.

The group G acts from the left on $X'(\bar{k})$, if we define the action of $g \in G$, on a point $spec(\bar{k}) \to X'$ by composing it from the left with $spec(g) : spec(\bar{k}) \to spec(\bar{k})$. If H is an open subgroup of G, then we can identify the set of fixed points $X'(\bar{k})^H$ with the set $X'(k')$ of all k'-valued points on X'. Here k' is the fixed field of H, hence finite over k, and the inclusion $X'(k') \subset X'(\bar{k})$ is induced by the canonical morphism $spec(\bar{k}) \to spec(k')$. G acts continuously on $X'(\bar{k})$, since $X'(\bar{k}) = \bigcup_H X'(\bar{k})^H$.

Let T_G denote the canonical topology on the category of all continuous left G-sets (cp. I, (1.3.3)). We have

(2.1) Theorem. *The functor $X' \to X'(\bar{k})$ is an equivalence of topologies between the étale site $spec(k)_{ét}$ of $spec(k)$ and the canonical topology T_G.*

Remark. The statement of the theorem means that $X' \to X'(\bar{k})$ is an equivalence of categories, and that both $X' \to X'(\bar{k})$ and any functor quasi-inverse to $X' \to X'(\bar{k})$ are morphisms of topologies.

Proof: Let us denote the functor $X' \mapsto X'(\bar{k})$ by f. First note that $(X' \times_{Z'} Y')(\bar{k}) \cong X'(\bar{k}) \times_{Z'(\bar{k})} Y'(\bar{k})$, hence f commutes with fibre products.

Let $\{X'_i \to X'\}$ be a family of morphisms of étale k-schemes. We show that $\{X'_i \to X'\}$ is a covering in $spec(k)_{ét}$ if and only if $\{X'_i(\bar{k}) \to X'(\bar{k})\}$ is a covering in T_G. Since both categories have arbitrary sums, and since f commutes with sums, it suffices to show that a morphism $Y' \to X'$ of étale k-schemes is surjective if and only if $Y'(\bar{k}) \to X'(\bar{k})$ is surjective. Assume then that $Y' \to X'$ is onto. Let $x' \in X'$ and let a k-homomorphism $k(x') \to \bar{k}$ be given. If $y' \in Y'$ lies above x', then the extension $k(y')/k(x')$ is finite and separable, and therefore $k(x') \to \bar{k}$ extends to a k-homomorphism $k(y') \to \bar{k}$. But this means that $y'(\bar{k}) \mapsto x'(\bar{k})$ is surjective. Conversely, assume that $y'(\bar{k}) \mapsto x'(\bar{k})$ is onto. Let $x' \in X'$. Since $k(x')/k$ is finite and separable, there is a $\bar{k}$-valued point corresponding to x'. If we take any $\bar{k}$-valued point of Y' lying above it, then the corresponding point $y' \in Y'$ lies above x'. Hence $Y' \to X'$ is surjective.

What we have shown so far proves the equivalence of topologies, provided we know that f is an equivalence of categories. To see this we first show the existence of the left adjoint functor ${}^{ad}f$ of f, and then verify that the adjoint morphisms are isomorphisms (cp. (0.1.1)).

To prove the existence of ${}^{ad}f$ it suffices to show that the functor

$$X' \mapsto \mathrm{Hom}_G(U, X'(\bar{k}))$$

is representable for all continuous G-sets U (cp. (0.1.1)).

Now, each continuous G-set is equal to the sum of its orbits, and each orbit is isomorphic to a continuous G-set of the form G/H for an open subgroup H of G. Therefore it is enough to show that the functors

$$X' \mapsto \mathrm{Hom}_G(G/H, X'(\bar{k}))$$

are representable, since the category of étale k-schemes has arbitrary sums.

Let k' be the fixed field of the open subgroup H. Then $spec(k')$ is an étale k-scheme, and we have the isomorphisms

$$\mathrm{Hom}_G(G/H, X'(\bar{k})) \cong X'(\bar{k})^H \cong X'(k') = \mathrm{Hom}_k(spec(k'), X'),$$

which are functorial in X'. Therefore $spec(k')$ represents the functor $X' \mapsto \mathrm{Hom}_G(G/H, X'(\bar{k}))$.

The adjoint map $G/H \to f({}^{ad}f(G/H)) = spec(k')(\bar{k})$ is the G-map, which sends the class $1 \cdot H$ to the $\bar{k}$-valued point $spec(\bar{k}) \to spec(k')$ corresponding to the inclusion $k' \subset \bar{k}$. But this map is an isomorphism. Since f and ${}^{ad}f$ commute with direct sums, we obtain $id \cong f \circ {}^{ad}f$. In a similar way we obtain $id \cong {}^{ad}f \circ f$.

This proves Theorem (2.1) (see also [16], exp. V, 4.) □

Remark. Theorem (2.1) implies in particular that the topology on $spec(k)_{\text{ét}}$ coincides with the canonical topology (I, (1.3.1)) on the category of étale k-schemes. This fact does not hold in general for arbitrary schemes.

(2.2) Corollary. *i) The functor*

$$F \mapsto \varinjlim F(spec\ k')$$

is an equivalence between the category of abelian sheaves on $spec(k)_{\text{ét}}$ and the category of continuous G-sets. Here k' runs through all finite (or only through all finite normal) extensions of k in $\bar{k}$.

ii) For all abelian sheaves F on $spec(k)_{\text{ét}}$ we have ∂-functorial isomorphisms

$$H^q_{\text{ét}}(spec(k), F) \cong H^q(G, \varinjlim F(spec(k'))).$$

The right-hand side denotes Galois-cohomology (cp. [32]).

This is immediate from (2.1) together with I, (1.3.3) and I, (3.3.3).

(2.3) Corollary. *Let k be separably closed. Then the section functor $F \mapsto F(spec(k))$ is an equivalence between the category of abelian sheaves on $spec(k)_{\acute{e}t}$ and the category $\mathcal{A}b$. For all sheaves F on $spec(k)_{\acute{e}t}$ we have $H^q(spec(k), F) = 0$ for $q > 0$.*

For the last statement see also example I, (3.5.4).

§ 3. Examples of Étale Sheaves

(3.1) Representable Sheaves

Let X be a scheme. By definition, a presheaf F of sets (of abelian groups) on $X_{\acute{e}t}$ is a **sheaf**, if for each covering $\{X'_i \to X'\}$ in $X_{\acute{e}t}$ the 'condition for sheaves' holds, i.e. the diagram

$$F(X') \to \prod F(X'_i) \rightrightarrows \prod F(X'_i \times_{X'} X'_j)$$

is exact (cp. I, (1.2.3)).

(3.1.1) Lemma. *For a presheaf F of sets (of abelian groups) on $X_{\acute{e}t}$ the following are equivalent:*

i) F is a sheaf.

ii) The 'condition for sheaves' holds for coverings in $X_{\acute{e}t}$ of the following types:

a) $\{X'_i \to X'\}$ is a surjective family of open immersions.

b) $\{Y' \to X'\}$ is a single surjective morphism of affine schemes.

Proof: i) $\Longrightarrow$ ii) holds trivially. To prove the implication ii) $\Longrightarrow$ i) we use the criterion of I, (3.1.4). Therefore it suffices to find a refinement $\{Y'_j \to X'\}$ of a given covering $\{X'_i \to X'\}$, which satisfies the 'condition for sheaves'.

Let $\{X'_i \to X'\}$ be an arbitrary covering in $X_{\acute{e}t}$. We choose an affine open covering

$$X' = \bigcup_j U'_j$$

of X'. This yields for each j a covering

$$\{X'_i \times_{X'} U'_j \to U'_j\}$$

in $X_{\text{ét}}$. Let us now cover each of the schemes $X'_i \times_{X'} U'_j$ by affine open subsets U'_{ijk}. We then obtain a refinement

$$\{U'_{ijk} \to U'_j\} \to \{X'_i \times_{X'} U'_j \to U'_j\}$$

of $\{X'_i \times_{X'} U'_j \to U'_j\}$. As an affine scheme, U'_j is quasi-compact. Moreover, as étale morphisms the maps $U'_{ijk} \to U'_j$ are open. Therefore we can refine the covering $\{U'_{ijk} \to U'_j\}$ by a finite subcovering, say $\{U'_{j\ell} \to U'_j\}$:

$$\{U'_{j\ell} \to U'_j\} \to \{U'_{ijk} \to U'_j\}.$$

If we compose the covering $\{U'_{j\ell} \to U'_j\}$ with the affine open covering $\{U'_j \to X'\}$, we obtain a covering

$$\{U'_{j\ell} \to X'\}$$

together with a natural refinement map

$$\{U'_{j\ell} \to X'\} \to \{X'_i \to X'\}.$$

But as we said at the beginning of the proof we only have to show that $\{U'_{j\ell} \to X'\}$ satisfies the 'condition for sheaves'. By construction, $\{U'_{j\ell} \to X'\}$ is the composite of $\{U'_j \to X'\}$ and $\{U'_{j\ell} \to U'_j\}$. Now

$$(*) \qquad \begin{cases} \{U'_j \to X'\} & \text{is a family of open immersions and} \\ \{U'_{j\ell} \to U'_j\} & \text{is a finite family of morphisms} \\ & \text{of affine schemes} \end{cases}$$

In general, if a presheaf satisfies the 'condition for sheaves' for a covering $\{U'_j \to X'\}$ and all coverings $\{U'_{j\ell} \to U'_j\}$, then it is obvious that it satisfies the condition also for the composite.

Hence we have finally reduced our problem to showing that the 'condition for sheaves' holds for coverings of the type mentioned in $(*)$.

For the covering $\{U'_j \to X'\}$ consisting of open immersions this is true by assumption a).

For a typical covering $\{U'_\ell \to U'\}$ given by a finite family of morphisms of affine schemes, we argue as follows:

We form the affine scheme $\coprod U_j'$, which allows us to write the covering $\{U_j' \to U'\}$ as a composite of the coverings

$$\{U_j' \to \coprod U_j'\} \text{ and } \{\coprod U_j' \to U'\}.$$

By a), the presheaf satisfies the 'condition for sheaves' for the first covering and by b) it satisfies it for the second, hence for the composite. □

Remark. It is possible to axiomatize the special topological properties of the site $X_{\text{ét}}$ used in the proof of (3.1.1). The lemma can then be extended to more general topologies with essentially the same proof (cp. [7], exp. IV, 6.2.).

(3.1.2) Theorem. *The coverings in $X_{\text{ét}}$ are families of universal effective epimorphisms (I, (1.3.1)) in the category of X-schemes. In other words: For each X-scheme Z the functor $X' \mapsto \mathrm{Hom}_X(X', Z)$ is a* **sheaf** *on $X_{\text{ét}}$.*

Remark. This implies in particular that the topology on $X_{\text{ét}}$ is coarser than the canonical topology on the category of étale X-schemes.

Proof: It suffices to show that the coverings of type a) and b) in (3.1.1) are families of universal effective epimorphisms in the category of X-schemes.

Given an open covering $X' = \bigcup X_i'$ a morphism $\varphi : X' \to Z$ is uniquely determined by its restrictions $\varphi_i : X_i' \to Z$. Conversely, for each system of morphisms $\varphi_i : X_i' \to Z$, such that $\varphi_i|_{X_i' \cap X_j'} = \varphi_j|_{X_i' \cap X_j'}$, there exists a unique morphism $\varphi : X' \to Z$, such that $\varphi|_{X_i'} = \varphi_i$.

It remains to study a surjective X-morphism $Y' \to X'$ of affine schemes, which are étale over X. Such a morphism is faithfully flat and quasi-compact, and therefore we can quote the following general result from descent theory to finish the proof (cp. [16], exp. VIII, 5.3. or [23], II, 2.1.): A faithfully flat and quasi-compact morphism of schemes is a universal effective epimorphism in the category of schemes. □

Remark. It is quite possible that different X-schemes represent the same sheaf on $X_{\text{ét}}$. E.g. if Z is an arbitrary k-scheme, the sheaf represented by Z on $spec(k)_{\text{ét}}$ is also represented by an **étale** k-scheme Z' (cp. Theorem

(2.1) and I, (1.3.3.2)). We obtain uniqueness if we restrict our attention to the category of étale X-schemes.

Let $f : Y \to X$ be a morphism of schemes, and let Z be a X-scheme. We obtain a canonical morphism

$$\mathrm{Hom}_X(\,\cdot\,, Z) \to f_*\mathrm{Hom}_Y(\,\cdot\,, Z \times_X Y)$$

of sheaves on $X_{\text{ét}}$ if we assign to each X-morphism $X' \to Z$ the Y-morphism $X' \times_X Y \to Z \times_X Y$. In general the adjoint morphism

$$f^*\mathrm{Hom}_X(\,\cdot\,, Z) \to \mathrm{Hom}_Y(\,\cdot\,, Z \times_X Y)$$

of sheaves on $Y_{\text{ét}}$ is not an isomorphism. However, we have:

(3.1.3) Proposition. *If Z is an étale X-scheme, the canonical morphism*

$$f^*\mathrm{Hom}_X(\,\cdot\,, Z) \to \mathrm{Hom}_Y(\,\cdot\,, Z \times_X Y)$$

is an isomorphism. Hence if the étale X-scheme Z represents the sheaf F on $X_{\text{ét}}$, then the étale Y-scheme $Z \times_X Y$ represents the inverse image f^*F.

Proof: By definition, $f^*\mathrm{Hom}_X(\,\cdot\,, Z)$ is the sheaf associated to the presheaf $F^{\cdot}\mathrm{Hom}_X(\,\cdot\,, Z)$. Since Z is étale over X, the presheaf $f^{\cdot}\mathrm{Hom}_X(\,\cdot\,, Z)$ gets identified by I, (2.3.3) with the presheaf on $Y_{\text{ét}}$ represented by $Z \times_X Y$. The claim now follows from (3.1.2). □

If G is a group scheme over X, then we denote by G_X the sheaf on $X_{\text{ét}}$ represented by G. G_X is a sheaf of groups on $X_{\text{ét}}$. For each étale X-scheme X' we have $G_X(X') = \mathrm{Hom}_X(X', G)$, the group of points of G with values in X'.

If G is a commutative group scheme on X, then G_X is an abelian sheaf on $X_{\text{ét}}$. We mention the following examples:

The additive group $(\mathbb{G}_a)_X$:

We have $\mathbb{G}_a = spec(\mathbb{Z}[t]) \times_{spec(\mathbb{Z})} X$, and for an étale X-scheme X' we obtain:

$$\begin{aligned}(\mathbb{G}_a)_X(X') &= \mathrm{Hom}_X(X', spec(\mathbb{Z}[t]) \times_{spec(\mathbb{Z})} X) \\ &= \mathrm{Hom}(X', spec(\mathbb{Z}[t])) \\ &= \mathrm{Hom}(\mathbb{Z}[t], \Gamma(X', \mathcal{O}_{X'})) \\ &= \Gamma(X', \mathcal{O}_{X'}).\end{aligned}$$

The multiplicative group $(\mathbb{G}_m)_X$:

We have $\mathbb{G}_m = spec(\mathbb{Z}[t, t^{-1}]) \times_{spec(\mathbb{Z})} X$, and for an étale X-scheme X' we obtain

$$\begin{aligned}(\mathbb{G}_m)_X(X') &= \mathrm{Hom}_X(X', spec(\mathbb{Z}[t, t^{-1}]) \times_{spec(\mathbb{Z})} X)) \\ &= \mathrm{Hom}(\mathbb{Z}[t, t^{-1}], \Gamma(X', \mathcal{O}_{X'})) \\ &= \Gamma(X', \mathcal{O}_{X'})^*.\end{aligned}$$

The sheaf $(\mu_n)_X$ of n-th roots of unity:

We have $\mu_n = spec(\mathbb{Z}[t]/(t^n - 1)) \times_{spec(\mathbb{Z})} X$, and for an étale X-scheme X' we obtain:

$$\begin{aligned}(\mu_n)_X(X') &= \mathrm{Hom}_X\big(X', spec(\mathbb{Z}[t]/(t^n - 1)) \times_{spec(\mathbb{Z})} X\big) \\ &= \mathrm{Hom}(\mathbb{Z}[t]/(t^n - 1), \Gamma(X', \mathcal{O}_{X'})) \\ &= \{s \in \Gamma(X', \mathcal{O}_{X'}) / s^n = 1\}.\end{aligned}$$

For each natural number n we have the following exact sequence of abelian sheaves on $X_{\text{ét}}$:

$$0 \to (\mu_n)_X \to (\mathbb{G}_m)_X \xrightarrow{n} (\mathbb{G}_m)_X,$$

where $(\mathbb{G}_m)_X \xrightarrow{n} (\mathbb{G}_m)_X$ denotes the n-th power morphism $s \mapsto s^n$. Under suitable assumptions on n with respect to the characteristics of the local residue fields of X, the map $(\mathbb{G}_m)_X \xrightarrow{n} (\mathbb{G}_m)_X$ is surjective (cp. § 4).

Let A be a discrete abelian group. We denote by A_X the sheaf, which is associated to the presheaf $X' \to A$ for étale X-schemes X'. A_X is called the **constant sheaf** with values in A. Obviously we have

$$\begin{aligned}A_X(X') &= \prod_{\text{connected components of } X'} A \\ &= \mathrm{Hom}_X(X', \textstyle\coprod_A X).\end{aligned}$$

This means that the constant sheaf A_X is represented by the étale group scheme $\amalg_A X$ with the group structure induced by A. If F is an arbitrary abelian sheaf on $X_{ét}$, each homomorphism $A \to F(X)$ induces a natural morphism of sheaves $A_X \to F$. In fact, this gives an isomorphism between $\mathrm{Hom}(A, F(X))$ and $\mathrm{Hom}(A_X, F)$ by I, (3.1).

Consider the constant sheaf $(\mathbb{Z}/n\mathbb{Z})_X$ on $X_{ét}$. The morphisms $(\mathbb{Z}/n\mathbb{Z})_X \to F$ from $(\mathbb{Z}/n\mathbb{Z})_X$ to an abelian sheaf F correspond uniquely to those sections of F over X, which are annihilated by n. This implies:

The isomorphisms $(\mathbb{Z}/n\mathbb{Z})_X \cong (\mu_n)_X$ of sheaves correspond uniquely to the **primitive n-th root of 1 on X**, hence to those sections of $(\mu_n)_X$ over X, which have order precisely n on each connected component of X. In particular, we see that the sheaf $(\mu_n)_X$ is isomorphic to the constant sheaf $(\mathbb{Z}/n\mathbb{Z})_X$ if and only if there exists at least one primitive root n-th of 1 on X. For example, $(\mu_2)_X \cong (\mathbb{Z}/2\mathbb{Z})_X$ in case $X = spec(\mathbb{Z})$. Note, however, that the defining group schemes are not isomorphic.

Let us assume now that n is relatively prime to the characteristics of all local residue fields of X. Obviously, this is equivalent to saying that n is invertible on X or that the group scheme $\mu_n = spec(\mathbb{Z}[t]/(t^n-1)) \times_{spec(\mathbb{Z})} X$ is étale over X. Then we have

(3.1.4) The sheaf $(\mu_n)_X$ is **locally isomorphic** to the sheaf $(\mathbb{Z}/n\mathbb{Z})_X$, in other words, there is a covering $\{X_i \to X\}$ in $X_{ét}$, such that the restrictions $(\mu_n)_X/X_i = (\mu_n)_{X_i}$ are isomorphic to $(\mathbb{Z}/n\mathbb{Z})_{X_i}$.

In fact, if $U = spec(A)$ is an affine open subset of X, the map $U' = spec(A[t]/(t^n - 1)) \to U$ is étale and surjective, and the class of $t \bmod (t^n - 1)$ is a primitive n-th root of 1 on U'.

(3.2) Étale Sheaves of $\mathcal{O}_X$-Modules

(3.2.1) Theorem. *Let M be a quasi-coherent sheaf of $\mathcal{O}_X$-modules on X. Then*

$$X' \mapsto \Gamma(X', M \otimes_{\mathcal{O}_X} \mathcal{O}_{X'})$$

is an abelian sheaf on $X_{ét}$, denoted by $M_{ét}$.

Proof: Again, we use lemma (3.1.1): In case a), i.e. in case of a Zariski-open covering of an étale X-scheme X', the 'condition for sheaves' is trivially satisfied by $M_{\text{ét}}$, since $M \otimes_{\mathcal{O}_X} \mathcal{O}_{X'}$ is a **sheaf** of $\mathcal{O}_{X'}$-modules.

Since M is quasi-coherent, case b) is a consequence of the following result (cp. [16], exp. VIII, 1.7.):

If $A \to B$ is faithfully flat and if M is an A-module, then the diagram

$$M \to M \otimes_A B \rightrightarrows M \otimes_A B \otimes_A B$$

is exact (see also [23], ch. II, 2.2.). □

The functor $M \to M_{\text{ét}}$ from the category of quasi-coherent $\mathcal{O}_X$-modules to the category of abelian sheaves on $X_{\text{ét}}$ is additive and left exact. This follows, since $M \to M \otimes_{\mathcal{O}_X} \mathcal{O}_{X'}$ is exact for flat morphisms $X' \to X$. Furthermore, $(\mathcal{O}_X)_{\text{ét}} = (\mathbb{G}_a)_X$, and for a locally-free $\mathcal{O}_X$-module M of rank r the sheaf $M_{\text{ét}}$ is locally isomorphic to $(\mathbb{G}_a)_X^r$ on $X_{\text{ét}}$. We will compute the cohomology groups $H^q(X, M_{\text{ét}})$ in § 4.

(3.3) Appendix: The Big Étale Site.

Let X be a scheme, and let Sch/X denote the category of X-schemes. We define a topology on Sch/X by taking as coverings surjective families $\{X'_i \to X'\}$ of **étale** X-morphisms $X'_i \to X'$. This topology is called the **big étale site of X.**

Since X-morphisms of étale schemes are étale (1.1.2), the inclusion functor $i : Et/X \to Sch/X$ of categories is in fact a morphism of topologies, and we obtain the functors (cp. I, (3.6.1))

$$res = i^s : \widetilde{Sch/X} \to \widetilde{X}_{\text{ét}}$$
$$ext = i_s : \widetilde{X}_{\text{ét}} \to \widetilde{Sch/X}$$

between the categories of abelian sheaves on Sch/X and $X_{\text{ét}}$. They are given as follows: If F is an abelian sheaf on Sch/X, we have $res\, F(X') = F(X')$ for étale X-schemes X'. If G is an abelian sheaf on $X_{\text{ét}}$, then the sheaf $ext(G)$ is associated to the presheaf

$$Z \mapsto \varinjlim G(X')$$

on Sch/X, where the limit extends over diagrams of the form

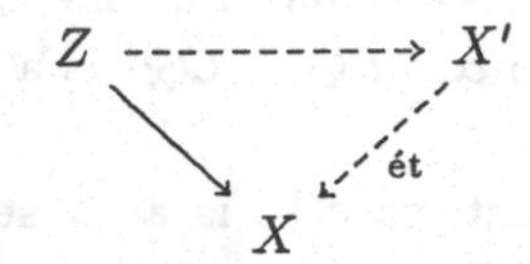

The functor ext is exact (cp. I, 3.6.7) and left adjoint to res.

(3.3.1) Theorem. *i) The adjoint morphism $G \to res(ext\, G)$ is an isomorphism for all abelian sheaves G on $X_{\text{ét}}$, hence the functor $ext : \widetilde{X}_{\text{ét}} \to \widetilde{Sch/X}$ is fully faithful. Moreover, the functor $res : \widetilde{Sch/X} \to \widetilde{X}_{\text{ét}}$ is exact.*

ii) There are canonical isomorphisms

$$H^p(X_{\text{ét}}; X, res\, F) \cong H^p(Sch/X; X, F)$$
$$H^p(X_{\text{ét}}; X, G) \cong H^p(Sch/X; X, ext\, G)$$

for all abelian sheaves F on Sch/X and all abelian sheaves G on $X_{\text{ét}}$.

Proof: This follows immediately from I, (3.9.2) and I, (3.9.3), since the conditions i) and ii), which are needed there, hold trivially. □

Remark. Because of this theorem it does not matter for the study of the cohomology of étale sheaves on X, whether one uses the 'small' étale site or the big étale site of X.

Besides the étale topology there are various other topologies, which can be introduced on the category Sch/X by using various restrictions on the morphisms $X_i' \to X'$ in coverings $\{X_i' \to X'\}$ (cp. [16], exp. IV, 6). It is worth mentioning the faithfully-flat quasi-compact topology on Sch/X (f.p.q.c. for short). It is defined as the finest topology, such that surjective families of open immersions and quasi-compact faithfully flat morphisms are coverings. The f.p.q.c. topology is finer than the étale topology. The sheaves defined in 3.1.2. and 3.3.1. are also sheaves on the f.p.q.c. topology.

§4. The Theories of Artin-Schreier and of Kummer

(4.1) The Groups $H^q(X, (\mathbb{G}_a)_X)$

Let X be a scheme. We know by (3.2) that the abelian sheaf $(\mathbb{G}_a)_X$ on $X_{\text{ét}}$ can be described as $(\mathbb{G}_a)_X = (\mathcal{O}_X)_{\text{ét}}$. In the following we will more generally compute the cohomology groups $H^q(X, M_{\text{ét}})$ for an arbitrary quasi-coherent sheaf of $\mathcal{O}_X$-modules M on X.

We use the relation between étale and Zariski cohomology described in (1.3.1.): If $\varepsilon : X_{Zar} \hookrightarrow X_{\text{ét}}$ denotes the natural inclusion, the Leray spectral sequence for the sheaf $M_{\text{ét}}$ takes the form

$$E_2^{pq} = H^p_{Zar}(X, R^q\varepsilon^s(M_{\text{ét}})) \Longrightarrow R^{p+q} = H^{p+q}_{\text{ét}}(X, M_{\text{ét}}). \tag{4.1.1}$$

Here $\varepsilon^s(M_{\text{ét}})$ is the sheaf $U \mapsto M_{\text{ét}}(U) = \Gamma(U, M)$ on X_{Zar}, hence $\varepsilon^s(M_{\text{ét}}) = M$.

(4.1.2) Theorem. *For each quasi-coherent sheaf M of $\mathcal{O}_X$-modules and for each p the edge morphisms*

$$H^p_{Zar}(X, M) \to H^p_{\text{ét}}(X, M_{\text{ét}})$$

of the spectral sequence (4.1.1) are isomorphisms. In particular,

$$H^p_{\text{ét}}(X, (\mathbb{G}_a)_X) \cong H^p_{Zar}(X, \mathcal{O}_X).$$

Serre's well known theorem on the Zariski cohomology of quasi-coherent sheaves of $\mathcal{O}_X$-modules on affine schemes (cp. [19], 1.3.1) then implies the following

(4.1.3) Corollary. *For an affine scheme we have*

$$H^p_{\text{ét}}(X, M_{\text{ét}}) = 0 \quad \textit{for} \quad p > 0,$$

in particular

$$H^p_{\text{ét}}(X, (\mathbb{G}_a)_X) = 0 \quad \textit{for} \quad p > 0.$$

Proof of (4.1.2): We show that $R^q\varepsilon^s(M_{\text{ét}}) = 0$ for $q > 0$, which implies the bijectivity of the edge morphisms $H^p_{Zar}(X, M) \to H^p_{\text{ét}}(X, M_{\text{ét}})$.

The sheaf $R^q\varepsilon^s(M_{\text{ét}})$ is the sheaf associated to the presheaf (cp. I, (3.7.1))

$$U \mapsto H^q(X_{\text{ét}}; U, M_{\text{ét}}).$$

By (1.4.9) we have $H^q(X_{\text{ét}}; U, M_{\text{ét}}) \cong H^q(U_{\text{ét}}; U, (M_{\text{ét}})/U)$, and – obviously – the restriction $(M_{\text{ét}})/U$ of $M_{\text{ét}}$ to U equals $(M/U)_{\text{ét}}$. Since the affine open subsets U of X form a basis of the topology X_{Zar}, it will be enough for the verification of $R^q\varepsilon^s(M_{\text{ét}}) = 0$ for $q > 0$ to prove Corollary (4.1.3).

Therefore assume X is affine. Let T be the full subcategory of $X_{\text{ét}}$ of all affine schemes endowed with the induced topology. The comparison theorem I, (3.9.1) shows that

$$H^q(X_{\text{ét}}; X; M_{\text{ét}}) \cong H^q(T; X, M_{\text{ét}}),$$

where we denoted the sheaf induced by $M_{\text{ét}}$ on T again by $M_{\text{ét}}$.

We claim that the sheaf $M_{\text{ét}}$ is flabby on T. By I, (3.5.3) this would indeed imply $H^q(T; X, M_{\text{ét}}) = 0$ for $q > 0$.

Obviously, the topology T is noetherian (I, (3.10.1)). By I, (3.10.2) $M_{\text{ét}}$ will therefore be flabby on T, if we can show that

$$H^q(\{Y_i \to Y\}, M_{\text{ét}}) = 0$$

for $q > 0$ and for all **finite** coverings $\{Y_i \to Y\}$ in T. For a finite covering $\{Y_i \to Y\}$ the scheme $\coprod_i Y_i$ is again affine, and furthermore

$$H^q(\{Y_i \to Y\}, M_{\text{ét}}) = H^q(\coprod_i Y_i \to Y, M_{\text{ét}})$$

for $q \geq 0$ by I, (1.2.3) and I, (2.2.3). Therefore we have reduced the proof to showing the following:

If $\{Z \to Y\}$ is a covering in T consisting only of one morphism, the cohomology groups $H^q(\{Z \to Y\}, M_{\text{ét}})$ vanish for $q > 0$, hence the complex of Čech cochains (cp. I, (2.2.2))

$$0 \to M_{\text{ét}}(Y) \to C^0(\{Z \to Y\}, M_{\text{ét}}) \xrightarrow{d} C^1(\{Z \to Y\}, M_{\text{ét}}) \xrightarrow{d} \cdots$$

is exact. Let $Y = spec(A)$ and let $Z = spec(B)$, hence M is an A-module. Then the complex is:

$$0 \to M \to M \otimes_A B \xrightarrow{d} M \otimes_A B \otimes_A B \xrightarrow{d} \cdots$$

Since B/A is faithfully flat, this complex – known as the **Amitsur complex** – is exact by descent theory (cp. [23], II, 2.1 or [15], 3.1.). □

(4.2) The Artin-Schreier Sequence

Let X be a scheme, whose characteristic is the prime number p, i.e. $p \cdot \mathcal{O}_{X,x} = 0$ for all points $x \in X$.

We consider the constant sheaf $(\mathbb{Z}/p\mathbb{Z})_X$ on $X_{\text{ét}}$. The unit section of $(\mathbb{G}_a)_X$ over X defines a morphism of sheaves

$$(\mathbb{Z}/p\mathbb{Z})_X \to (\mathbb{G}_a)_X.$$

This morphism is injective, since the unit section of $(\mathbb{G}_a)_X$ over X has order precisely p.

Furthermore, we have the morphism of sheaves

$$F : (\mathbb{G}_a)_X \to (\mathbb{G}_a)_X,$$

defined by $s \to s^p$ for $s \in (\mathbb{G}_a)_X(X') = \Gamma(X', \mathcal{O}_{X'})$, X' an étale X-scheme. F is called **Frobenius morphism**. We define

$$\wp = id - F : (\mathbb{G}_a)_X \to (\mathbb{G}_a)_X$$

and prove:

(4.2.1) Theorem. *The sequence*

$$0 \to (\mathbb{Z}/p\mathbb{Z})_X \to (\mathbb{G}_a)_X \xrightarrow{\wp} (\mathbb{G}_a)_X \to 0$$

of morphisms of abelian sheaves on $X_{\text{ét}}$ is exact. It is called the **Artin-Schreier-sequence** *on X.*

Proof: Let X' be an étale X-scheme and $s \in ker(\wp)(X')$. Then $s \in \Gamma(X', \mathcal{O}_{X'})$, $s = s^p$, hence s is local and constant on each connected component of X'. In other words, s is a section of $im((\mathbb{Z}/p\mathbb{Z})_X \to (\mathbb{G}_a)_X)$ over X'.

Again, let X' be an étale X-scheme, and let $s \in (G_a)_X(X') = \Gamma(X', \mathcal{O}_{X'})$. To prove the surjectivity of $\wp : (\mathbb{G}_a)_X \to (\mathbb{G}_a)_X$, we have to show that there exists a covering $\{X'_i \to X'\}$ of X' in $X_{\text{ét}}$, such that each of the induced sections $s_i \in (\mathbb{G}_a)_X(X'_i) = \Gamma(X'_i, \mathcal{O}_{X'_i})$ is of the form $s_i = t_i^p - t_i = \wp(t_i)$ for some $t_i \in \Gamma(X'_i, \mathcal{O}_{X'_i})$ (cp. I, (3.2)). But this is clear from the following observation:

Let A be a ring of characteristic p and let $s \in A$. The A-algebra $A[t]/(t - t^p - s)$ is free of rank p and étale over A. Moreover, the class $t' = t \bmod (t - t^p - s)$ satisfies $\wp(t') = s$. □

The Artin-Schreier-sequence (4.2.1) yields the following long exact sequence in cohomology (cp. (4.1.2)):

$$\begin{aligned} 0 &\to H^0(X, (\mathbb{Z}/p\mathbb{Z})_X) \to H^0(X, \mathcal{O}_X) \to H^0(X, \mathcal{O}_X) \\ &\to H^1(X, (\mathbb{Z}/p\mathbb{Z})_X) \to H^1(X, \mathcal{O}_X) \to H^1(X, \mathcal{O}_X) \\ &\to H^2(X, (\mathbb{Z}/p\mathbb{Z})_X) \to \cdots . \end{aligned}$$

We obtain:

(4.2.2) Corollary. *If X has characteristic p, then there is an exact sequence*

$$0 \to H^0(X, \mathcal{O}_X)/\wp H^0(X, \mathcal{O}_X) \to H^1(X, (\mathbb{Z}/p\mathbb{Z})_X) \to H^1(X, \mathcal{O}_X)^F \to 0,$$

where $H^1(X, \mathcal{O}_X)^F$ denotes the subgroup of $H^1(X, \mathcal{O}_X)$ of elements fixed under the Frobenius endomorphism F.

(4.2.3) Corollary. *If $X = spec(A)$ is affine of characteristic p, then*

$$H^q(X, (\mathbb{Z}/p\mathbb{Z})_X) = \begin{cases} A/\wp A & \text{if } q = 1 \\ 0 & \text{if } q > 1. \end{cases}$$

(4.2.4) Corollary. *Assume k is a separably closed field of characteristic p and X is a reduced proper k-scheme. Then*

$$H^1(X, (\mathbb{Z}/p\mathbb{Z})_X) \cong H^1(X, \mathcal{O}_X)^F.$$

(4.3) The Groups $H^q(X, (\mathbb{G}_m)_X)$

Let X be a scheme. By definition (cp. (3.1)) we have

$$H^0(X, (\mathbb{G}_m)_X) = \Gamma(X, \mathcal{O}_X^*).$$

In dimension 1 the following holds:

(4.3.1) Theorem *(Hilbert's Theorem 90). There is a canonical isomorphism*

$$H^1(X, (\mathbb{G}_m)_X) \cong Pic(X),$$

where $Pic(X)$ *denotes the* **Picard group of** X, *i.e. the group of isomorphism classes of invertible* $\mathcal{O}_X$*-modules on* X *(cp. [17],* 0_I*, 5.6.).*

Proof: Let $\varepsilon : X_{Zar} \to X_{\text{ét}}$ denote the natural inclusion of topologies. By (1.3.1) we have the following spectral sequence for the abelian sheaf $(\mathbb{G}_m)_X$ on $X_{\text{ét}}$:

$$E_2^{pq} = H^p_{Zar}(X, R^q\varepsilon^s(\mathbb{G}_m))_X \Longrightarrow E^{p+q} = H^{p+q}_{\text{ét}}(X, (\mathbb{G}_m)_X).$$

Now $\varepsilon^s(\mathbb{G}_m)_X = \mathcal{O}_X^*$, and by [17], 0_I, 5.6.3 the group $Pic(X)$ is canonically isomorphic to $H^1_{Zar}(X, \mathcal{O}_X^*)$. Therefore (4.3.1) can be restated more precisely as: The edge morphism

$$H^1_{Zar}(X, \mathcal{O}_X^*) \to H^1_{\text{ét}}(X, (\mathbb{G}_m)_X$$

of the above spectral sequence is an isomorphism.

This edge morphism is part of the exact sequence of terms of low degree (cp. 0, (2.3.3))

$$0 \to H^1_{Zar}(X, \mathcal{O}_X^*) \to H^1_{\text{ét}}(X, (\mathbb{G}_m)_X \to H^0_{Zar}(X, R^1\varepsilon^s(\mathbb{G}_m)_X) \to \cdots$$

Therefore the bijectivity of the edge morphism is a consequence of the following

(4.3.2) Lemma. $R^1\varepsilon^s(\mathbb{G}_m)_X = 0$.

In fact, this lemma is equivalent to (4.3.1): $R^1\varepsilon^s(\mathbb{G}_m)_X$ is the sheaf associated to the presheaf

$$U \mapsto H^1(X_{\text{ét}}; U, (\mathbb{G}_m)_X) = H^1_{\text{ét}}(U, (\mathbb{G}_m)_U)$$

on X_{Zar} (cp. (1.4.8), (1.4.9)). If we assume that $H^1_{\text{ét}}(U, (\mathbb{G}_m)_U) \cong Pic(U)$ for all open U in X, the associated sheaf, however, is the zero sheaf, since to any given invertible $\mathcal{O}_U$-module L there exists an open covering of U trivializing L. (In general, the considerations in I, (3.1) show that the sheaf $F^\#$ associated to an abelian presheaf F on a topology T is zero if and only if for each $U \in T$ and each $s \in F(U)$ there exists a covering $\{U_i \to U\}$, such that $s \mapsto 0$ under $F(U) \to \prod F(U_i)$.)

Since the affine open subsets of X form a basis for the Zariski topology, we see that it suffices for the proof of (4.3.2) to prove (4.3.1) for **affine** schemes.

Hence assume X to be affine. By I, (3.4.7) we have

$$H^1_{\text{ét}}(X, (\mathbb{G}_m)_X) \cong \check{H}^1(X, (\mathbb{G}_m)_X),$$

and by definition (cp. I, (2.2.5))

$$\check{H}^1(X, (\mathbb{G}_m)_X) = \varinjlim_{\{X_i \to X\}} H^1(\{X_i \to X\}, (\mathbb{G}_m)_X),$$

where $\{X_i \to X\}$ runs through all coverings of X in $X_{\text{ét}}$. Since X is affine, we can refine each covering of X by a finite covering $\{X_i \to X\}$ with affine schemes X_i (cp. I, (2.2.4)). Moreover,

$$H^1(\{X_i \to X\}, (\mathbb{G}_m)_X) = H^1(\coprod_i X_i \to X, (\mathbb{G}_m)_X),$$

and therefore

$$H^1_{\text{ét}}(X, (\mathbb{G}_m)_X) \cong \varinjlim H^1(\{Y \to X\}, (\mathbb{G}_m)_X),$$

where now Y runs through all surjective affine étale morphisms with target X.

By [23], V, 2.1. there is a canonical isomorphism

$$H^1(\{Y \to X\}, (\mathbb{G}_m)_X) \cong ker(Pic(X) \to Pic(Y)).$$

Since for any invertible $\mathcal{O}_X$-module L there exists an affine étale faithfully flat X-scheme Y trivializing L (we can choose a finite affine covering $X = \bigcup U_i$ with $L/U_i \cong \mathcal{O}_X/U_i$ and take $Y = \coprod U_i$), we have

$$\varinjlim ker(Pic(X) \to Pic(Y)) = Pic(X)$$

and hence in fact

$$H^1_{\text{ét}}(X, (\mathbb{G}_m)_X) \cong Pic(X). \qquad \square$$

Remark. For $q \geq 2$ the edge morphisms

$$H^q_{Zar}(X, \mathcal{O}_X^*) \to H^q_{\text{ét}}(X, (\mathbb{G}_m)_X)$$

are in general not isomorphisms. If we take e.g. for X the spectrum of a field k, then we always have $H^q_{Zar}(X, \mathcal{O}_X^*) = 0$ for $q \geq 1$, whereas the group $H^2_{\text{ét}}(X, (\mathbb{G}_m)_X)$, which we can identify by (2.2) and Galois cohomology with the Brauer group of k (cp. [32], II, 1.2.), is in general non-trivial.

(4.3.3) Remark. The group $H^2(X, (\mathbb{G}_m)_X)$ is closely related to the Brauer group $Br(X)$ of the scheme X, which is the group of equivalence classes of Azumaya algebras on X. More precisely, we have an injective homomorphism

$$\delta : Br(X) \to H^2(X, (\mathbb{G}_m)_X).$$

An **Azumaya algebra** on X is a sheaf A of $\mathcal{O}_X$-algebras, which is locally free of finite rank as an $\mathcal{O}_X$-module and satisfies the following two equivalent conditions:

i) For each point $x \in X$ the algebra $A_x \otimes_{\mathcal{O}_{X,x}} k(x)$ is a central simple $k(x)$-algebra.

ii) The canonical homomorphism

$$A \otimes_{\mathcal{O}_X} A^{op} \to \mathcal{E}nd_{\mathcal{O}_X}(A)$$

is an isomorphism.

Two Azumaya algebras A and A' on X are called **equivalent**, if there are locally free $\mathcal{O}_X$-modules E and E' of finite rank, such that

$$A \otimes_{\mathcal{O}_X} \mathcal{E}nd_{\mathcal{O}_X}(E) \cong A' \otimes_{\mathcal{O}_X} \mathcal{E}nd_{\mathcal{O}_X}(E').$$

The equivalence classes of Azumaya algebras on X form an abelian group with multiplication induced by the tensor product, the so-called **Brauer group** $Br(X)$ of X.

For more on the construction of the injective map $\delta : Br(X) \to H^2(X, (\mathbb{G}_m)_X)$, on conditions for the surjectivity of δ and on the computation of $Br(X)$ the reader may consult [12], exp. IV, V, VI or [11].

(4.4) The Kummer Sequence

Let X be a scheme. For each natural number n, raising to the n-th power defines a morphism $(\mathbb{G}_m)_X \xrightarrow{n} (\mathbb{G}_m)_X$, and we have an exact sequence

$$0 \to (\mu_n)_X \to (\mathbb{G}_m)_X \xrightarrow{n} (\mathbb{G}_m)_X.$$

(4.4.1) Theorem. *Let n be invertible on X, i.e. n is prime to $char(k(x))$ for all $x \in X$. Then there is an exact sequence*

$$0 \to (\mu_n)_X \to (\mathbb{G}_m)_X \xrightarrow{n} (\mathbb{G}_m)_X \to 0$$

of morphisms of abelian sheaves on $X_{ét}$. This sequence is called the **Kummer sequence** *on X.*

Proof: We only have to show the surjectivity of $(\mathbb{G}_m)_X \xrightarrow{n} (\mathbb{G}_m)_X$. The proof is similar to the proof of the surjectivity in the Artin-Schreier-sequence (4.2.1): For an étale X-scheme X' and $s \in (\mathbb{G}_m)_X(X') = \Gamma(X', \mathcal{O}^*_{X'})$ we have to find a covering $\{X'_i \to X'\}$ of X' in $X_{ét}$ such that the maps $s_i \in (\mathbb{G}_m)_X(X'_i) = \Gamma(X'_i, \mathcal{O}^*_{X'_i})$, induced by s, are n-th powers in $\Gamma(X'_i, \mathcal{O}^*_{X'_i})$.

This is a consequence of the following observation: Let A be a ring with n invertible in A, and let $s \in A^*$. Then the A-algebra $A[t]/(t^n - s)$ is free of rank n and étale over A. Moreover, for the class $t' = t \bmod (t^n - s)$ we have $t'^n = s$. □

If n is invertible on X, the Kummer sequence yields the long exact sequence

$$\begin{aligned} 0 &\to H^0(X, (\mu_n)_X) \to H^0(X, \mathcal{O}^*_X) \xrightarrow{n} H^0(X, \mathcal{O}^*_X) \\ &\to H^1(X, (\mu_n)_X) \to Pic(X) \xrightarrow{n} Pic(X) \\ &\to H^2(X, (\mu_n)_X) \to \cdots \end{aligned}$$

For an abelian group A let us denote by ${}_nA$ and A_n the kernel and the cokernel of the endomorphism $a \mapsto n \cdot a$ of A. With these notations we obtain:

(4.4.2) Corollary. *If n is invertible on X, then the sequence*

$$0 \to H^0(X, \mathcal{O}^*_X)_n \to H^1(X, (\mu_n)_X) \to {}_nPic(X) \to 0$$

is exact.

(4.4.3) Corollary. *If X is the spectrum of a local ring A, and if n is invertible in A, then we have*

$$H^1(X, (\mu_n)_X) \cong A^*/A^{*n}.$$

(4.4.4) Corollary. *Let X be a reduced proper scheme over a separably closed field k, and let $char(k)$ be prime to n. Then we have*

$$H^1(X,(\mu_n)_X) \cong {}_nPic(X).$$

Remark. For one-dimensional schemes X we will consider the groups $H^q(X,(\mu_n)_X)$ for $q \geq 2$ more closely in section (10.3).

(4.5) The Sheaf of Divisors on $X_{ét}$

Let X be a noetherian scheme, and let K denote the ring of rational functions on X. K is an artinian ring, whose local components are the local rings $\mathcal{O}_{X,x_i}$ of X at the maximal points x_i of X (cp. [17], 8.1.7.). Let $j : spec(K) \to X$ be the canonical morphism. It induces a natural morphism of abelian sheaves on $X_{ét}$ (cp. (3.1)):

$$(\mathbb{G}_m)_X \to j_*(\mathbb{G}_m)_K.$$

Here we have used the abbreviation $(\mathbb{G}_m)_K$ for the sheaf $(\mathbb{G}_m)_{spec(K)}$ on $spec(K)_{ét}$.

Let us now assume that the scheme X has **no embedded components** (cp. [20], 3.1.). This implies that the morphism $(\mathbb{G}_m)_X \to j_*(\mathbb{G}_m)_K$ is **injective**, since $j : spec(K) \to X$ is dominant ([17], 5.4.), and hence also $spec(K) \times_X X' \to X'$ for each étale X-scheme X'. We define an abelian sheaf $\mathcal{D}iv_X$ on $X_{ét}$ via the exact sequence

$$0 \to (\mathbb{G}_m)_X \to j_*(\mathbb{G}_m)_K \to \mathcal{D}iv_X \to 0.$$

This sheaf is called the **sheaf of divisors on** $X_{ét}$.

For the computation of $H^0(X,\mathcal{D}iv_X)$ we apply the left exact functor ε^s to the above exact sequence, where $\varepsilon : X_{Zar} \to X_{ét}$. We obtain the following exact sequence of morphisms of abelian sheaves on X_{Zar}:

$$0 \to \mathcal{O}_X^* \to \mathcal{K}_X^* \to \varepsilon^s(\mathcal{D}iv_X) \to R^1\varepsilon^s(\mathbb{G}_m)_X \to \cdots$$

Here $\mathcal{K}_X$ is the sheaf of rational functions on X ([17], 8.3). By (4.3.2) we have $R^1\varepsilon^s(\mathbb{G}_m)_X = 0$. Hence the sheaf $\varepsilon^s(\mathcal{D}iv_X)$ gets identified with the sheaf of divisors in the Zariski topology, and therefore $H^0(X,\mathcal{D}iv_X)$ is equal to the group of divisors on X_{Zar} ([20], 2.1.1.).

If X' is an étale X-scheme of finite type, then X' is noetherian and has no embedded components by [20], 3.3.3. Hence for the restriction of $\mathcal{D}iv_X$ to X' we obtain $\mathcal{D}iv_X/X' = \mathcal{D}iv_{X'}$.

As described in [20], 21.6. there is a canonical morphism of sheaves

$$cyc : \mathcal{D}iv_X \to \bigoplus_{x \in X^{(1)}} (i_x)_* \mathbb{Z}_x.$$

Here $X^{(1)}$ denotes the set of points $x \in X$ with $\dim \mathcal{O}_{X,x} = 1$, $i_x : spec(k(x)) \to X$ denotes the canonical morphism, and $\mathbb{Z}_x$ denotes the constant sheaf on $spec(k(x))_{ét}$ with value $\mathbb{Z}$.

If we assume now X to be regular, then we see from [20], 21.6.9 that $cyc : \mathcal{D}iv_X \to \bigoplus_{x \in X^{(1)}} (i_x)_* \mathbb{Z}_x$ is an isomorphism.

As a noetherian scheme, X is quasi-compact and quasi-separable (cp. [17], 6.1.13.), hence we obtain from (1.5.3):

$$H^1(X, \mathcal{D}iv_X) \cong \bigoplus_{x \in X^{(1)}} H^1(X, (i_x)_* \mathbb{Z}_x) = 0,$$

where the vanishing of the right-hand side is a consequence of the following more general

(4.5.1) Lemma. *Let X be a scheme, $x \in X$, and let $i : spec(k(x)) \to X$ be the canonical morphism. Then*

$$H^1(X, i_* \mathbb{Z}_x) = 0.$$

Proof: The Leray spectral sequence (1.4.3)

$$H^p(X, R^q i_*(\mathbb{Z}_x)) \Longrightarrow H^{p+q}(spec(k(x)), \mathbb{Z}_x)$$

induces an injection (cp. 0, (2.3.3))

$$H^1(X, i_* \mathbb{Z}_x) \hookrightarrow H^1(spec(k(x)), \mathbb{Z}_x).$$

By (2.2) the right-hand side gets identified with $H^1(G, \mathbb{Z})$, where the Galois-group G of $\overline{k(x)}/k(x)$ acts trivially on $\mathbb{Z}$. By Galois-cohomology (cp. [32], I, 2-3.) the group $H^1(G, \mathbb{Z})$ is equal to the group of **continuous** homomorphisms from G to $\mathbb{Z}$, hence trivial. □

Remark. The statement of lemma (4.5.1) remains true if $\mathbb{Z}$ is replaced by an arbitrary torsionfree abelian group.

The result $H^1(X, \mathcal{D}iv_X) = 0$ gives information about the "cohomological Brauer group" $H^2(X, (\mathbb{G}_m)_X)$ of X. Before we consider this, we need one more lemma:

(4.5.2) Lemma. *Let X be a scheme, $x \in X$, and let $i : spec(k(x)) \to X$ denote the canonical morphism. Then $R^1 i_*(\mathbb{G}_m)_{k(x)} = 0$, and therefore the canonical homomorphism*

$$H^2(X, i_*(\mathbb{G}_m)_{k(x)}) \to H^2(spec(k(x)), (\mathbb{G}_m)_{k(x)})$$

is injective.

(Here we used again the abbreviation $(\mathbb{G}_m)_{k(x)}$ for $(\mathbb{G}_m)_{spec(k(x))}$.)

Proof: $R^1 i_*(\mathbb{G}_m)_{k(x)}$ is the sheaf associated to the presheaf $X' \mapsto H^1(X' \times_X spec(k(x)), (\mathbb{G}_m)_{k(x)})$. But this presheaf is the zero sheaf, since by Hilbert's Theorem 90 the group $H^1(u, (\mathbb{G}_m)_{k(x)})$ vanishes in general for all étale $k(x)$-schemes u. The other statement follows from the exact sequence of terms of low degree

$$0 \to E_2^{1,0} \to E^1 \to E_2^{0,1} \to E_2^{2,0} \to E^2$$

resulting from the Leray spectral sequence

$$H^p(X, R^q i_*(\mathbb{G}_m)_{k(x)}) \Longrightarrow H^{p+q}(spec(k(x)), (\mathbb{G}_m)_{k(x)}).$$ □

(4.5.3) Proposition. *Let X be a regular noetherian scheme. There is a canonical injective homomorphism*

$$H^2(X, (\mathbb{G}_m)_X) \hookrightarrow \prod Br(K_i),$$

where K_i runs through the fields of rational functions on the finitely many irreducible components of X.

Proof: From the exact sequence

$$0 \to (\mathbb{G}_m)_X \to j_*(\mathbb{G}_m)_K \to \mathcal{D}iv_X \to 0,$$

and the fact that $H^1(X, \mathcal{D}iv_X) = 0$, we see that

$$H^2(X, (\mathbb{G}_m)_X) \to H^2(X, j_*(\mathbb{G}_m)_K)$$

is injective. Moreover by Lemma (4.5.2) we have the injection

$$H^2(X, j_*(\mathbb{G}_m)_K) \to \prod H^2(spec(K_i), (\mathbb{G}_m)_{K_i}) = \prod Br(K_i). \qquad \square$$

(4.5.4) Corollary. *Let X be a regular algebraic curve over a separably closed field k. Then $H^2(X, (\mathbb{G}_m)_X) = 0$. For all n prime to the characteristic of k there is an exact sequence*

$$\begin{aligned} 0 &\to H^0(X, (\mu_n)_X) \to H^0(X, \mathcal{O}_X^*) \xrightarrow{n} H^0(X, \mathcal{O}_X^*) \\ &\to H^1(X, (\mu_n)_X) \to Pic(X) \xrightarrow{n} Pic(X) \\ &\to H^2(X, (\mu_n)_X) \to 0. \end{aligned}$$

Proof: The statement $H^2(X, (\mathbb{G}_m)_X) = 0$ follows from (4.5.3) together with Tsen's Theorem, which says that $Br(K) = 0$ for field extensions K of transcendence degree 1 over a separably closed field k (cp. [32], II, § 3). The exact sequence simply follows from the Kummer sequence (4.4.1). $\square$

Remark. Later we will see that the sequence in (4.5.4) remains exact for algebraic curves with singularities as well (cp. (10.3.4)).

§ 5. Stalks of Étale Sheaves

Let X be a scheme. A **geometric point of** X is defined to be an X-scheme of the form $P = spec(\Omega)$ with Ω a separably closed field. Equivalently, a geometric point of X is a point $x \in X$ together with an injection of $k(x)$ into a separably closed field Ω.

For a scheme $P = spec(\Omega)$ with Ω a separably closed field we know by (2.3) that the section functor $F \mapsto F(P)$ is an equivalence between the category of abelian sheaves on $P_{ét}$ and the category $\mathcal{A}b$ of abelian groups.

(5.1) Definition. *Let P be a geometric point of X and let $u : P \to X$ be its structure morphism. For an abelian sheaf F on $X_{ét}$ the abelian group*

$$F_P = u^*F(P)$$

is called the **stalk of** F **in** P.

Example: Let G be an étale commutative group scheme over X and let G_X denote the abelian sheaf on $X_{\text{ét}}$ represented by G. For a geometric point $u : P = spec(\Omega) \to X$ the sheaf u^*G_X is then represented by the P-group scheme $G \times_X P$ (cp. (3.1.3)). We obtain

$$\begin{aligned} u^*G_X(P) &= \mathrm{Hom}_P(P, G \times_X P) \\ &= \mathrm{Hom}_X(P, G), \end{aligned}$$

which means that the stalk $G_{X,P}$ of G_X in $P = spec(\Omega)$ is equal to the group of Ω-valued points of G. For example, the stalk $(\mu_n)_{X,P}$, for n invertible on X, is equal to the group of n-th roots of 1 in Ω, and the stalk $A_{X,P}$ of the constant abelian sheaf A_X on $X_{\text{ét}}$, defined by a discrete abelian group A, is equal to A.

(5.2) Proposition. *i) The functor $F \mapsto F_P$ is exact and commutes with inductive limits.*

ii) If $v : P' \to P$ is an X-morphism of geometric points of X, we have

$$F_{P'} \cong F_P.$$

iii) If $f : X \to Y$ is a morphism of schemes and if P is a geometric point of X, then P is naturally a geometric point of Y as well. Furthermore, for each abelian sheaf F on $Y_{\text{ét}}$ we have

$$(f^*F)_P \cong F_P.$$

Proof: i) Let $u : P \to X$ be the structure morphism of P. By definition, $F \mapsto F_P$ is the composite of the functors $u^* : \widetilde{X}_{\text{ét}} \to \widetilde{P}_{\text{ét}}$ and $\Gamma_P : \widetilde{P}_{\text{ét}} \to Ab$. Both functors are exact and commute with inductive limits (cp. (1.4.2), iii) and (2.3)).

ii) For the structure morphisms $u' : P' \to X$ and $u : P \to X$ we have $u' = u \circ v$. For an abelian sheaf F on $X_{\text{ét}}$ we obtain

$$F_{P'} = u'^*F(P') = (v^*u^*F)(P') \cong u^*F(P) = F_P,$$

since $v_{\text{ét}} : P_{\text{ét}} \to P'_{\text{ét}}$ is an equivalence of categories, and hence $G(P) \cong v^*G(P')$ for all abelian sheaves G on $P_{\text{ét}}$.

iii) Let u be the structure morphism of P. Then as a geometric point of Y the structure morphism of P is given by $f \circ u : P \to Y$. Hence $(f^*F)_P \cong F_P$, since $(f \circ u)^* = u^*f^*$. □

(5.3) Remark. Let us choose a separable closure $\overline{k(x)}$ of the residue field $k(x)$ of a point $x \in X$. Let $\bar{x}$ denote the geometric point of X corresponding to $\overline{k(x)}$. We can reformulate statement ii) of (5.2) as follows: The stalks F_P of all geometric points P of X, which are localized in x, are (non-canonically) isomorphic to the stalk $F_{\bar{x}}$.

Again, let $u : P \to X$ be a geometric point of X and let F be an abelian sheaf on $X_{\text{ét}}$. In order to compute the stalk F_P we consider the **category of étale neighbourhoods of P in X**. This category is also called the category of P-punctured étale X-schemes and consists of pairs (X', u') with X' an étale X-scheme and $u' : P \to X'$ an X-morphism. By (1.4.1) the dual category of this category is filtered, and for the presheaf $u^{\cdot}F$ of F under $u : P \to X$ we obtain at P:

$$u^{\cdot}F(P) = \varinjlim_{X'} F(X'),$$

where the inductive limit extends over the dual category of the category of étale neighbourhoods of P in X. Now u^*F is the sheaf associated to the presheaf $u^{\cdot}F$, and we obtain a canonical homomorphism

$$\varinjlim_{X'} F(X') \to F_P.$$

In particular, for each étale neighbourhood X' of P we have a canonical homomorphism $F(X') \to F_P$. Given $s \in F(X')$, we denote the image of s in F_P by s_P.

(5.4) Proposition. *The canonical homomorphism*

$$\varinjlim_{X'} F(X') \to F_P$$

is an isomorphism for all abelian sheaves F on $X_{\text{ét}}$. Even more generally:

$$\varinjlim_{X'} G(X') \to (G^{\#})_P$$

is an isomorphism for all abelian presheaves G on $X_{\text{ét}}$.

Proof: To prove the first isomorphism it suffices to show that the canonical map $G(P) \to G^{\#}(P)$ is an isomorphism for any abelian presheaf G on $P_{\text{ét}}$. By construction of $G^{\#}$ (cp. I, (3.1.1)) we have to show that

$$G(P) \to \check{H}^0(P, G) = \varinjlim_{\{U_i \to P\}} H^0(\{U_i \to P\}, G)$$

is an isomorphism. Now, for each covering $\{U_i \to P\}$ in $P_{\text{ét}}$ there is a refinement map (I, (2.2.4))

$$\{P \xrightarrow{id} P\} \to \{U_i \to P\}.$$

This already proves that $G(P) \to \check{H}^0(P, G)$ is an isomorphism.

It also proves the second isomorphism in the proposition, if we take into account the following

(5.4.1) Lemma. *Let $f : X \to Y$ be a morphism of schemes, and let G be an abelian presheaf on $Y_{\text{ét}}$. Then the canonical morphism*

$$(f^{\cdot}G)^{\#} \to f^*(G^{\#}),$$

induced by $f^{\cdot}G \to f^{\cdot}(G^{\#}) \to f^*(G^{\#})$, is an isomorphism.

Proof: More generally, $f : T \to T'$ be a morphism of topologies and let G be an abelian presheaf on T. Then $(f_P G)^{\#} \to f_s(G^{\#})$ is an isomorphism. This is a consequence of the canonical isomorphism

$$\text{Hom}((f_P G)^{\#}, G') \cong \text{Hom}(f_s(G^{\#}), G')$$

for all abelian sheaves G' on T', which is easily obtained from I, (2.3), (3.1), (3.6). □

Example: Let $f : X \to Y$ be a morphism of schemes, F an abelian sheaf on $X_{\text{ét}}$, and let P be a geometric point of Y. Then we have

$$R^q f_*(F)_P \cong \varinjlim_{Y'} H^q(X \times_Y Y', F), \tag{5.5}$$

where Y' runs through the étale neighbourhoods of P in Y.

Example: Let k be a field, $\bar{k}$ be a separable closure of k and let $G = \text{Gal}(\bar{k}/k)$. As was shown in § 2, we have an equivalence between the category of abelian sheaves on $spec(k)_{\text{ét}}$ and the category of continuous G-modules given by

$$F \mapsto \varinjlim_{k'} F(spec(k')).$$

Here k' runs through all finite subextensions of $\bar{k}/k$.

Viewing $spec(\bar{k})$ as a geometric point of $spec(k)$ allows us to identity this functor with the stalk functor $F \to F_{spec(\bar{k})}$. In fact, $spec(k')$ together with the morphism $spec(\bar{k}) \to spec(k')$ induced by $k' \subset \bar{k}$ is an étale neighbourhood of $spec(\bar{k})$ for all finite subextensions k' of $\bar{k}/k$. The full subcategory of these étale neighbourhoods clearly is initial in the category of all étale neighbourhoods of $spec(\bar{k})$. Therefore by (5.4) and 0, (3.3.1) we have

$$\varinjlim_{k'} F(spec(k')) \cong F_{spec(\bar{k})}.$$

As in (5.3) we choose a separable closure $\overline{k(x)}$ of the residue field $k(x)$ of a point $x \in X$, where X is an arbitrary scheme. Again let $\bar{x}$ denote the geometric point of X corresponding to $\overline{k(x)}$. As we observed in (5.3), it is sufficient for the study of the stalks of an abelian sheaf F on $X_{\text{ét}}$ to look at the stalks $F_{\bar{x}}$. Note that each of these stalks $F_{\bar{x}}$ has a natural structure of a G_x-module, where G_x is the Galois group of $\overline{k(x)}$ over $k(x)$.

We can now prove the following result about the family of all stalk functors $F \mapsto F_{\bar{x}}$ $(x \in X)$:

(5.6) Theorem. *i) A morphism $F' \to F$ of abelian sheaves on $X_{\text{ét}}$ is an isomorphism (monomorphism, epimorphism), if and only if the morphism $F'_{\bar{x}} \to F_{\bar{x}}$ is an isomorphism (monomorphism, epimorphism) for all $x \in X$.*

ii) A morphism $v : F' \to F$ of abelian sheaves on $X_{\text{ét}}$ is 0 if and only if $v_{\bar{x}} = 0$ for all $x \in X$. In particular, for $s \in F(X)$ we have $s = 0$ if and only if $s_{\bar{x}} = 0$ for all $x \in X$.

iii) A sequence $F' \to F \to F''$ of morphisms of abelian sheaves on $X_{\text{ét}}$ is exact if and only if the sequence $F'_{\bar{x}} \to F_{\bar{x}} \to F''_{\bar{x}}$ is exact for all $x \in X$.

Proof: Let us first show that the statement about the isomorphism in part i) implies all the other statements of the theorem: By (5.2) i) the functor $F \mapsto F_{\bar{x}}$ is exact and hence commutes with kernels and cokernels. Now $v : F' \to F$ is a monomorphism if and only if $F' \to im(v)$ is an isomorphism, and $v : F' \to F$ is an epimorphism if and only if $im(v) \to F$ is an isomorphism. Finally, $v : F' \to F$ is the zero morphism if and only if $o \to im(v)$ is an isomorphism. Hence we obtain the remaining statements in i) and the first statement in ii). Given $s \in F(X)$, we have $s = 0$ if and only if the induced morphism of sheaves $\mathbb{Z}_X \to F$ is the zero morphism.

Since this morphism equals $1 \mapsto s_{\bar{x}}$ on each stalk (note that $\mathbb{Z}_{X,\bar{x}} = \mathbb{Z}$), we obtain the second statement in ii). To prove iii), assume we have a sequence $F' \xrightarrow{v} F \xrightarrow{u} F''$ of morphisms of abelian sheaves, which is exact at each stalk. Then we have $(uv)_{\bar{x}} = u_{\bar{x}}v_{\bar{x}} = 0$, hence $uv = 0$ and therefore $im(v) \to ker(u)$ is an isomorphism.

We turn now to the proof of the key statement in i): Assume $v : F' \to F$ is a morphism of abelian sheaves on $X_{\acute{e}t}$, such that $v_{\bar{x}} : F'_{\bar{x}} \to F_{\bar{x}}$ is an isomorphism for each $x \in X$. We want to show that v is an isomorphism.

We show first that v is a monomorphism. Therefore let X' be an étale X-scheme and let $s \in F'(X')$, such that $v(s) = 0$. Looking at the restriction (cp. (1.4.8)) of $v : F' \to F$ to X' and using parts ii) and iii) of (5.2), we may as well assume that $X' = X$. Then $0 = v(s)_{\bar{x}} = v_{\bar{x}}(s_{\bar{x}})$, hence by assumption $s_{\bar{x}} = 0$ for all $x \in X$. Now by (5.1) and the fact that the dual category of the category of étale neighbourhoods of $\bar{x}$ is filtered, we find an étale neighbourhood X'_x of $\bar{x}$, such that s maps to 0 under $F'(X) \to F'(X'_x)$. But $\{X'_x \to X\}_{x \in X}$ is a covering in $X_{\acute{e}t}$ and therefore we obtain $s = 0$, since F' is a sheaf.

To prove surjectivity we show the following: Given an étale X-scheme X' and $t \in F(X')$, there is $s \in F'(X')$, such that $v(s) = t$. Again, we may assume without loss of generality that $X' = X$. Then each $v_{\bar{x}} : F'_{\bar{x}} \to F_{\bar{x}}$ is onto and similar considerations as above in the proof of injectivity show that we can find an étale neighbourhood X'_x of $\bar{x}$ and a section $s_x \in F'(X'_x)$, such that $v(s_x)$ equals the image of t under $F(X) \to F(X'_x)$. For two points x and y of X the restrictions of s_x and s_y to $X'_x \times_X X'_y$ coincide, since we know already that

$$F'(X'_x \times_X X'_y) \to F(X'_x \times_X X'_y)$$

is injective. Therefore the family $(s_x)_{x \in X}$ defines a section $s \in F'(X)$ which maps to t under v by construction. □

Let us keep the notations of (6.5). Given an abelian sheaf F on $X_{\acute{e}t}$ and $s \in F(X)$, the subset

$$supp(s) = \{x \in X \mid s_{\bar{x}} \neq 0\}$$

of X is called the **support of** s. $supp(s)$ is a Zariski-closed subset of X: In fact, if $s_{\bar{x}} = 0$, we find an étale neighbourhood $\varphi : X' \to X$ of $\bar{x}$

such that the restriction of s vanishes on X'. This is done as in the proof of (5.6). But then $s_{\bar{y}} = 0$ for all y from $\varphi(X')$, which is a Zariski-open neighbourhood of x in X.

The subset

$$\{x \in X \mid F_{\bar{x}} \neq (0)\}$$

of X is not necessarily Zariski-closed. Its closure is called the **support of the sheaf** F, denoted by $supp(F)$. Since by (5.6) the vanishing of a sheaf is equivalent to the vanishing of all its stalks, we see that $supp(F)$ is equal to the complement of the largest open subset U of X, such that the restriction F/U vanishes.

§ 6. Strict Localizations

(6.1) Henselian Rings and Strictly Local Rings

In this section we summarize for later use results about henselian rings and strictly local rings. For a more detailed account of the theory we refer to [20], 18.5. – 18.8. or to [24].

(6.1.1) Theorem *(cp. [20], 18.5.11. – 18.5.13.). Let A be a local ring with maximal ideal $\mathfrak{m}$ and residue field $k = A/\mathfrak{m}$. The following statements are equivalent:*

i) Hensel's Lemma holds for A. This means: Given a normalized polynomial $f \in A[t]$, such that the reduction $\bar{f} \in k[t]$ factors as $\bar{f} = \bar{g} \cdot \bar{h}$ with normalized relatively prime polynomials $\bar{g}$ and $\bar{h}$ in $k[t]$, there exist polynomials $g, h \in A[t]$ representing $\bar{g}$ and $\bar{h}$, such that $f = g \cdot h$.

ii) For each finite A-algebra B the canonical A-homomorphism $B \to \prod_{\mathfrak{n}} B_{\mathfrak{n}}$ is an isomorphism. Here $\mathfrak{n}$ runs through the finitely many maximal ideals of B.

iii) Let B be an A-algebra. If B is the localization of a finitely generated A-algebra C at a prime ideal of C above $\mathfrak{m}$, and if $B/\mathfrak{m}B$ is finite over $A/\mathfrak{m} = k$, then B is finite over A.

iv) Let B be an A-algebra. If B is the localization of an étale A-algebra C at a prime ideal of C above $\mathfrak{m}$, and if $k \to B/\mathfrak{m}B$ is an isomorphism, then B is finite over A, and hence $A \to B$ is an isomorphism.

(6.1.2) Definition. *A local ring A satisfying the equivalent conditions i) – iv) of (6.1.1) is called* **henselian.**

Remark. Condition iv) in Theorem (6.1.1) has a more concrete geometric formulation:

iv') Let $S = spec(A)$ and let s denote the closed point of S. If X is an étale S-scheme and if x is a $k(s)$-rational point of the closed fibre $X_s = X \times_S k(s)$, there exists a (unique) section $u : S \to X$, such that $u(s) = x$.

To prove the equivalence of iv) and iv') note that the existence of a section $u : S \to X$ with $u(s) = x$ is the same as the bijectivity of $A \to \mathcal{O}_{X,x}$.

Example: If A is a local ring, which is separable and complete with respect to the $\mathfrak{m}$-adic topology, then A is henselian (cp. [20], 18.5.14).

Let A be a local ring with maximal ideal $\mathfrak{m}$, and let B be a finite étale A-algebra. Then $B/\mathfrak{m}B$ is a finite étale k-algebra, hence a finite separable k-algebra. We have the following fundamental result:

(6.1.4) Theorem (cp. [20], 18.5.15). *Let A be a henselian local ring. Then the functor $B \mapsto B/\mathfrak{m}B$ is an equivalence between the category of finite étale A-algebras and the category of finite separable k-algebras.*

Let A be a given local ring. We are looking for a local homomorphism $A \to A^h$ from A to a henselian local ring A^h, which is universal for local homomorphisms from A to henselian rings. This means that

$$\mathrm{Hom}_{loc}(A^h, B) \to \mathrm{Hom}_{loc}(A, B)$$

is bijective for all henselian rings B. This universal problem is in fact solvable: Define

$$A^h = \varinjlim B, \tag{6.1.5}$$

where B runs through the filtered category of essentially étale local A-algebras with trivial residue field extension (cp. [20], 18.6.). Here an A-algebra B is called **essentially étale**, if B is A-isomorphic to the localization of an étale A-algebra C at a prime ideal of C above $\mathfrak{m}$. A^h is called the **henselization** of A.

(6.1.6) Proposition. *Let A be a local ring and let A^h be its henselization. Then*

i) A^h is a faithfully flat A-module, $\mathfrak{m}A^h$ is the maximal ideal of A^h, and the homomorphism $A/\mathfrak{m} \to A^h/\mathfrak{m}A^h$ between the residue fields is bijective.

ii) A^h is noetherian if and only if A is noetherian.

For more properties of the henselization see [20], 18.6.6. – 18.6.14.

(6.1.7) Proposition. *Let A be a local ring. The following conditions are equivalent:*

i) A is henselian and the residue field is separably closed.

ii) A is henselian and all finite étale A-algebras are trivial, i.e. isomorphic to $A \times \cdots \times A$.

iii) If $A \to B$ is a local homomorphism from A to the localization B of an étale A-algebra, then $A \to B$ is an isomorphism.

iii') If $X \to S = spec(A)$ is étale and if x is a point of the closed fibre X_s, there exists a section $u : S \to X$, such that $u(s) = x$.

For the simple proof of this proposition see [20], 18.8.1.

(6.1.8) Definition. *A local ring satisfying the equivalent conditions i) – iii') of (6.1.7) is called* **strictly local**.

Let A be a given local ring with residue field k. Assume there is also a given homomorphism $k \to \Omega$ of k to a separably closed field Ω. We consider the category of essentially étale local A-algebras B together with a k-homomorphism $k(B) \to \Omega$ of the residue field $k(B)$ of B to Ω. Morphisms are local A-homomorphisms $B \to B'$ such that the diagram

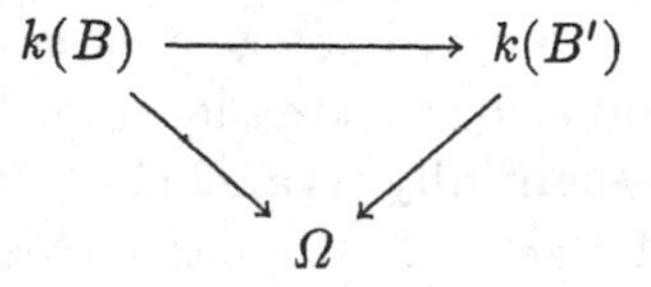

commutes. Now we define

$$A^{hs} = \varinjlim B.$$

(6.1.9) Proposition *(cp. [20], 18.8.8.). i) A^{hs} is a strictly local ring, whose residue field $k(A^{hs})$ can be identified with the separable closure of k in Ω.*

ii) A^{hs} is faithfully flat over A and $\mathfrak{m}A^{hs}$ is the maximal ideal of A^{hs}.

iii) A^{hs} is noetherian if and only if A is noetherian.

iv) The local homomorphism $A \to A^{hs}$ together with the k-homomorphism $k(A^{hs}) \to \Omega$ is universal for local homomorphisms $A \to A'$, where A' is strictly local and equipped with a k-homomorphism $k(A') \to \Omega$.

A^{hs} is called the **strict henselization of A with respect to $k \to \Omega$.**

(6.2) Strict Localization of a Scheme

Let X be a scheme and let $P = spec(\Omega) \to X$ be a geometric point of X. We consider the stalk of the abelian sheaf $(\mathcal{O}_X)_{\text{ét}} = (\mathbb{G}_a)_X$ on $X_{\text{ét}}$ in P, which we denote by $\mathcal{O}_{X,P}$ for simplicity. By (5.4) we have

$$\mathcal{O}_{X,P} = \varinjlim_{X'} \Gamma(X', \mathcal{O}_{X'}),$$

where X' runs through the filtered category, which is dual to the category of étale neighbourhoods of P in X. Let P be localized at $x \in X$. For a given étale neighbourhood X' of P we denote by x' the image of $P \to X'$. Then we get $\mathcal{O}_{X',x'} = \varinjlim \Gamma(U', \mathcal{O}_{X'})$, where U' runs through all Zariski-open neighbourhoods of x' in X'. All these U' are naturally étale neighbourhoods of P in X, and therefore

$$\mathcal{O}_{X,P} = \varinjlim \mathcal{O}_{X',x'},$$

where $\mathcal{O}_{X',x'}$ runs through the local rings of points x' in étale x-schemes X' above x, equipped with a $k(x)$-homomorphism $k(x') \to \Omega$. Hence by (6.1.9) $\mathcal{O}_{X,P}$ **is the strict henselization of $\mathcal{O}_{X,x}$ with respect to** $k(x) \to \Omega$.

(6.2.1) Definition. *$\mathcal{O}_{X,P}$ is called the* **strictly local ring of X in the geometric point P**. *The scheme $X(P) = spec(\mathcal{O}_{X,P})$ is called the* **strict localization of X in P**. *It is canonically an X-scheme with the structure morphism*

$$spec(\mathcal{O}_{X,P}) \to spec(\mathcal{O}_{X,x}) \to X.$$

The notion of strict localization of X in P allows the following interpretation of the stalk F_P of an arbitrary abelian sheaf F on $X_{\text{ét}}$:

(6.2.2) Proposition. *Let $F(P)$ denote the inverse image of F under $X(P) \to X$. There is a functorial isomorphism*

$$F_P \cong H^0(X(P), F(P)).$$

Proof: The structure morphism $P \to X$ of the geometric point P factors uniquely as $P \to X(P) \to X$. By means of (5.2) ii) and iii) the statement of (6.2.2) results from the following

(6.2.3) Lemma. *Let X be a strictly local scheme, i.e. X is the scheme of a strictly local ring. Let x be the closed point of X viewed as a geometric point of X. Then there is a functorial isomorphism*

$$F_x \cong H^0(X, F)$$

for all abelian sheaves F on $X_{\text{ét}}$, and $H^q(X, F) = 0$ for $q > 0$.

Proof: Let X' be an étale neighbourhood of x, and let $x' \in X'$ be the image of x. By (6.1.7), iii') there exists a section $u : X \to X'$, such that $u(x) = x'$. Therefore the étale neighbourhood X of x is initial in the category of all étale neighbourhoods of x. Hence by (5.4) there is an isomorphism $F_x \cong H^0(X, F)$, functorial in F. Since the stalk functor is exact by (5.2) i), we obtain $H^q(X, F) = 0$ for $q > 0$. □

(6.3) Étale Cohomology on Projective Limits of Schemes

Let I be a filtered category (0, (3.2)). We consider a contravariant functor from the category I to the category $\mathcal{S}ch$ of schemes:

$$\begin{cases} I^0 & \to \mathcal{S}ch \\ i & \mapsto X_i \end{cases}$$

We want to show that the projective limit (0, (3.1))

$$X = \varprojlim X_i$$

of this functor exists, provided that we assume the morphisms $X_j \to X_i$ to be affine for each arrow $i \to j$ in I. The scheme X is constructed as follows (cp. [20], 8.2.):

Let $i_0 \in I$ be fixed. For each object of the category I/i_0 of i_0-objects of I, e.g. for each arow $i \to i_0$, there is an X_{i_0}-isomorphism

$$X_i \cong spec(\mathcal{A}_i),$$

where $\mathcal{A}_i$ is a quasi-coherent sheaf of $\mathcal{O}_{X_{i_0}}$-algebras. For each arrow $i \to j$ in I/i_0 the X_{i_0}-morphism $X_j \to X_i$ induces a homomorphism $\mathcal{A}_i \to \mathcal{A}_j$ of $\mathcal{O}_{X_{i_0}}$-algebras. Therefore we obtain a functor $i \to \mathcal{A}_i$ from the category I/i_0, which is obviously again filtered, to the category of quasi-coherent sheaves of $\mathcal{O}_{X_{i_0}}$-algebras. Now for this functor the inductive limit $\mathcal{A} = \varinjlim \mathcal{A}_i$ exists, and we define

$$X = spec(\mathcal{A}).$$

The morphisms $\mathcal{A}_i \to \mathcal{A}$ induce morphisms $u_i : X \to X_i$, at first only for i_0-objects i, but then because of the filtration on I also for all $i \in I$. Together with these morphisms, X is the projective limit of the functor $i \mapsto X_i$. We note the following properties:

i) The underlying topological space of X is canonically homeomorphic to the projective limit of the underlying topological spaces of the X_i (cp. [20], 8.2.9.). Moreover, the structure sheaf $\mathcal{O}_X$ of X is canonically isomorphic to $\varinjlim u_i^{-1}(\mathcal{O}_{X_i})$ (cp. [20], 8.2.12.).

ii) If S is a scheme, and if all the X_i are S-schemes and the morphisms $X_j \to X_i$ are S-morphisms, then $X = \varprojlim X_i$ is also the projective limit of the X_i in the category of S-schemes (cp. [20], 8.2.3.). Moreover, if a morphism $T \to S$ is given, we have a canonical isomorphism $\varprojlim(X_i \times_S T) \cong X \times_S T$ (cp. [20], 8.2.5.).

(6.3.1) Example. Let X be a scheme and let $P \to X$ be a geometric point of X. Within the category of étale neighbourhoods of P in X we consider the full subcategory of **affine** étale neighbourhoods of P. This subcategory is initial, and we obtain for the strictly local ring of X in P (cp. (6.2.1)):

$$\mathcal{O}_{X,P} = \varinjlim_{X'} \Gamma(X', \mathcal{O}_{X'}),$$

where now X' runs through the dual category of the subcategory of affine étale neighbourhoods of P in X. This dual category is filtered, since the dual category of the category of all étale neighbourhoods of P has this property. Therefore we obtain

$$spec(\mathcal{O}_{X,P}) = \varprojlim X'.$$

This means that the strict localization $X(P) = spec(\mathcal{O}_{X,P})$ of X in P (cp. (6.2.1)) is equal to the projective limit of all affine étale neighbourhoods of P.

Let us return to the more general situation and assume that for each $i \in I$ we have an abelian sheaf F_i on $(X_i)_{ét}$ in such a way that for each arrow $i \to j$ in I the sheaf F_j is the inverse image of F_i under the corresponding morphism $X_j \to X_i$. If we define

$$F = u_i^* F_i,$$

this is an abelian sheaf on the étale site of $X = \varprojlim X_i$, independent of i. For each i we have a canonical homomorphism (cp. (1.4.4))

$$H^q(X_i, F_i) \to H^q(X, F),$$

which is compatible with the canonical homomorphism $H^q(X_i, F_i) \to H^q(X_j, F_j)$ induced by the arrow $i \to j$ in I (cp. (1.4.4)). Therefore we obtain a canonical homomorphism

$$\varinjlim H^q(X_i, F_i) \to H^q(X, F).$$

(6.3.2) Theorem. *Let I be a filtered category and let $i \mapsto X_i$ be a contravariant functor from the category I to the category of schemes. Assume that for each arrow $i \to j$ in I the morphisms $X_j \to X_i$ are affine (hence $X = \varprojlim X_i$ exists) and that all X_i are* **quasi-compact** *and* **quasi-separated.** *Assume further that for each $i \in I$ an abelian sheaf F_i on $(X_i)_{ét}$ is given, such that for $i \to j$ the sheaf F_j equals the inverse* *image of F_i under $X_j \to X_i$. Let F be the inverse image of F_i under $X \to X_i$, which is independent of i. Then the canonical homomorphism*

$$\varinjlim H^q(X_i, F_i) \to H^q(X, F)$$

is an isomorphism for all $q \geq 0$.

We do not want to give the rather technical proof of this theorem. We refer to [2], exp. VII, 5, exp. VI, 8.1. – 8.7. and [1], III, 3, where other versions of this theorem are discussed as well. We will use it in the following section for the interpretation of the stalks of $R^q f_* F$ by means of strict localizations, and then again in § 7 on the Artin spectral sequence.

(6.4) The Stalks of $R^q f_*(F)$.

Let $f : X \to Y$ be a morphism of schemes and let F be an abelian sheaf on $X_{\text{ét}}$. Let us consider the stalks $R^q f_*(F)_{\bar{y}}$ of the sheaf $R^q f_*(F)$ for points $y \in Y$. We know that

$$R^q f_*(F)_{\bar{y}} = \varinjlim_{Y'} H^q(X \times_Y Y', F)$$

(cp. (5.5)), where Y' runs through the dual category of the category of étale neighbourhoods of $\bar{y}$. In fact, we can restrict to affine Y' (cp. (6.3.1)), since arbitrary étale neighbourhoods of $\bar{y}$ are dominated by affine neighbourhoods.

Let $\bar{Y} = Y(\bar{y})$ be the strict localization of Y in $\bar{y}$ (cp. (6.2.1)). Then $\bar{Y} = \varprojlim Y'$, where Y' runs through the category of affine étale neighbourhoods of $\bar{y}$ (cp. (6.3.1)). Let us define

$$\bar{X} = X \times_Y \bar{Y} = \varprojlim X \times_Y Y',$$

and let $\bar{F}$ denote the inverse image of F under $\bar{X} \to X$. From (1.4.9) and the results in (6.3) we obtain a canonical homomorphism

$$R^q f_*(F)_{\bar{y}} = \varinjlim_{Y'} H^q(X \times_Y Y', F) \to H^q(\bar{X}, \bar{F}).$$

If we assume now that $f : X \to Y$ is quasi-compact and quasi-separated, the same is true for the morphisms $X \times_Y Y' \to Y'$. Since the schemes Y'

are affine, this implies that the schemes $X \times_Y Y'$ are all quasi-compact and quasi-separated. Therefore we obtain from (6.3.2):

(6.4.1) Theorem. *Let $f : X \to Y$ be a quasi-compact and quasi-separated morphism of schemes and let F be an abelian sheaf on $X_{\text{ét}}$. Furthermore, for $y \in Y$, let $\bar{Y}$ denote the strict localization in $\bar{y}$, let $\bar{X} = X \times_Y \bar{Y}$, and let $\bar{F}$ denote the inverse image of F under $\bar{X} \to X$. Then the canonical homomorphism*

$$R^q f_*(F)_{\bar{y}} \to H^q(\bar{X}, \bar{F})$$

is an isomorphism for all $q \geq 0$.

Remark. This theorem reduces the computation of the stalks of $R^q f_*(F)$ to the computation of the étale cohomology groups of those schemes, which are schemes over strict local rings, provided of course that the morphism $f : X \to Y$ is quasi-compact and quasi-separated.

As a first application of (6.4.1) we prove the following:

(6.4.2) Theorem. *Let $f : X \to Y$ be a* **finite** *morphism of schemes, and let F be an abelian sheaf on $X_{\text{ét}}$. Then we have:*

i) For each point $y \in Y$ there is a canonical isomorphism

$$(f_*F)_{\bar{y}} \cong \prod_{\bar{x} \in X_{\bar{y}}} F_{\bar{x}},$$

where $X_{\bar{y}}$ denotes the geometric fibre of $f : X \to Y$ in $\bar{y}$. In particular, f_*F commutes with arbitrary base change $Y' \to Y$ (cp. (1.4.7)).

ii) $R^q f_*(F) = 0$ for $q > 0$. Hence in particular the canonical homomorphisms (cp. (1.4.4))

$$H^q(Y, f_*F) \to H^q(X, F)$$

are isomorphic for $q \geq 0$.

Proof: By (6.4.1) we have an isomorphism

$$R^q f_*(F)_{\bar{y}} \cong H^q(\bar{X}, \bar{F}).$$

Since $X \to Y$ is a finite morphism, the same is true for $\bar{X} \to \bar{Y}$, and hence the main part of the theorem will follow from

(6.4.3) Lemma. *Let $X \to Y$ be finite and let Y be strictly local with closed point y. Let F be an abelian sheaf on $X_{\acute{e}t}$. Then we have:*

$$\begin{cases} H^0(X,F) & \cong \prod_{x\in X_y} F_x \\ H^q(X,F) & = 0 \quad \text{for } q > 0. \end{cases}$$

Proof of (6.4.3): Since $X \to Y$ is finite and Y is strictly local, the result in (6.1.1) ii) implies that $X = \coprod_x spec(\mathcal{O}_{X,x})$, where x runs through the finitely many closed points of X. These are precisely the points x of the closed fibre X_y. Each of these local rings $\mathcal{O}_{X,x}$ is henselian as a finite local $\mathcal{O}_{Y,y}$-algebra (cp. (6.1.1) ii)), and the residue field $k(x)$ of $\mathcal{O}_{X,x}$ is separably closed as a finite algebraic extension of $k(y)$. In other words, each of the local rings $\mathcal{O}_{X,x}$ for $x \in X_y$ is strictly local. Therefore we can apply lemma (6.2.3), which describes the cohomology of strictly local schemes, and we obtain:

$$\begin{cases} H^0(X,F) = \prod_{x\in X_y} H^0(spec(\mathcal{O}_{X,x}),F) \cong \prod_{x\in X_y} F_x \\ H^q(X,F) = \prod_{x\in X_y} H^q(spec(\mathcal{O}_{X,x}),F) = 0 \quad \text{for } q > 0. \quad \square \end{cases}$$

In order to obtain the formula in (6.4.2) i) from (6.4.3) we observe that $X_{\bar{y}} = \bar{X}_{\bar{y}}$. Each point $\bar{x} \in \bar{X}_{\bar{y}}$ is canonically a geometric point of X via the projection onto the factor X, and furthermore by (5.2) iii) we have $\bar{F}_{\bar{x}} = F_{\bar{x}}$.

Finally we have to show that f_*F commutes with arbitrary base change. This means (cp. (1.4.7)), that for a given cartesian diagram

$$\begin{array}{ccc} X' & \xrightarrow{f'} & Y' \\ {\scriptstyle g'}\downarrow & & \downarrow{\scriptstyle g} \\ X & \xrightarrow{f} & Y \end{array}$$

the base change morphism

$$g^*(f_*F) \to f'_*(g'^*F)$$

is an isomorphism. By (5.6) it suffices to check that the maps on the stalks

$$(g^*(f_*F))_{\bar{y}'} \to (f'_*(g'^*F))_{\bar{y}'}$$

are isomorphisms for all $y' \in Y'$. Let $y = g(y')$, and let $\bar{y}$ denote the geometric point of Y defined by the separable closure of $k(y)$ in $k(\bar{y}')$. We obtain

$$(g^*(f_*F))_{\bar{y}'} \cong (f_*F)_{\bar{y}'} \cong (f_*F)_{\bar{y}} \cong \prod_{\bar{x} \in X_{\bar{y}}} F_{\bar{x}}$$
$$(f'_*(g'^*F))_{\bar{y}'} \cong \prod_{\bar{x}' \in X'_{\bar{y}'}} (g'^*F)_{\bar{x}'}.$$

The scheme $X'_{\bar{y}'}$ results from the $k(\bar{y})$-scheme $X_{\bar{y}}$ by base change with the field $k(\bar{y}')$, and the canonical morphism $X'_{\bar{y}'} \to X_{\bar{y}}$ is a homeomorphism of the underlying topological spaces. If under this morphism $\bar{x}'$ is mapped to $\bar{x}$, we have $(g'^*F)_{\bar{x}'} \cong F_{\bar{x}'} \cong F_{\bar{x}}$. This finishes the proof of Theorem (6.4.2). □

Remark. More generally the following result is true: If $f : X \to Y$ is an **integral** morphism of schemes, and if F is an abelian sheaf on $X_{\text{ét}}$, then $R^q f_*(F) = 0$ for $q > 0$, and f_* commutes with arbitrary base change (cp. [2], exp. VIII, 5.6).

§ 7. The Artin Spectral Sequence

Let G be a profinite group, and let T_G be the canonical topology on the category of continuous left G-sets (cp. I, (1.3.3)). We want to consider the Leray spectral sequence for morphisms of topologies $\varphi : T_G \to T$.

We know that the category of abelian sheaves on T_G is equivalent to the category of continuous left G-modules (cp. I, (1.3.3.2)). Given a sheaf F on T, the sheaf $\varphi^s F$ on T_G corresponds under this equivalence to the continuous G-module $\varinjlim F(\varphi(G/H))$, where H runs through the open normal subgroups of G. If we look at an injective resolution of the sheaf F and use the fact that the functor $\varinjlim_H$ is exact (cp. 0, (3.2.1)), we see that more generally the sheaf $R^q\varphi^s(F)$ on T_G corresponds to the continuous G-module $\varinjlim H^q(T; \varphi(G/H), F)$. In view of this and the computations in I,

(3.3.3), we see that the Leray spectral sequence (I, (3.7.6)) for $\varphi : T_G \to T$ and the one-element G-set e takes the following form:

$$(7.1) \qquad H^p(G, \varinjlim H^q(T; \varphi(G/H), F)) \Longrightarrow H^{p+q}(T; \varphi(e), F).$$

In the special case that $\varphi : T_G \to X_{\text{ét}}$ is a morphism from T_G to the étale site of a scheme, this spectral sequence is also known as the **Artin spectral sequence** (cp. [1], III, 4.7.).

If we replace T_G by the canonical topology T'_G on the category of **finite** continuous left G-sets, the above considerations remain unchanged, and we obtain the same spectral sequence (cp. I, (3.9.4)).

Let X be a quasi-compact and quasi-separated scheme over a field k (e.g. an algebraic k-scheme). Let $\bar{k}$ denote a separable closure of k and let $\bar{X} = X \times_k \bar{k}$. As an application of the spectral sequence (7.1) we study the relations between the cohomology groups $H^q(X, F)$ of an abelian sheaf F on $X_{\text{ét}}$ and the groups $H^q(\bar{X}, \bar{F})$ of the inverse image $\bar{F}$ of F under $\bar{X} \to X$.

Let G be the Galois group of $\bar{k}/k$. If we compose the morphism $T_G \to spec(k)_{\text{ét}}$ with the morphism $spec(k)_{\text{ét}} \to X_{\text{ét}}$ induced from the structure morpism $X \to spec(k)$, we obtain a morphism

$$\varphi : T_G \to X_{\text{ét}}.$$

For an open normal subgroup H with fixed field k' we have $\varphi(G/H) = X \times_k k'$. In particular, $\varphi(e) = X$.

To compute the term $\varinjlim H^q(X_{\text{ét}}; \varphi(G/H), F)$ we use theorem (6.3.2). Obviously we have

$$\varprojlim \varphi(G/H) = \varprojlim X \times_k k' = X \times_k \bar{k} = \bar{X}.$$

The schemes $\varphi(G/H) = X \times_k k'$ are quasi-compact and quasi-separated, and therefore

$$\varinjlim H^q(X_{\text{ét}}; \varphi(G/H), F) \cong H^q(\bar{X}, \bar{F}).$$

We have shown:

(7.2) Theorem. *Let X be a quasi-compact and quasi-separated scheme over a field k, let $\bar{k}$ be a separable closure of k, $\bar{X} = X \times_k \bar{k}$, and let $G = Gal(\bar{k}/k)$. For any abelian sheaf F on $X_{\acute{e}t}$, we have a spectral sequence*

$$E_2^{pq} = H^p(G, H^q(\bar{X}, \bar{F})) \Longrightarrow E^{p+q} = H^{p+q}(X, F),$$

where $\bar{F}$ denotes the inverse image of F under $\bar{X} \to X$.

This spectral sequence describes the relationship between the cohomology groups of F and $\bar{F}$. For small dimensions we obtain the following exact sequence (cp. 0, (2.3.3)):

$$\begin{aligned} 0 \to H^1(G, H^0(\bar{X}, \bar{F})) &\to H^1(X, F) \to H^1(\bar{X}, \bar{F})^G \\ &\to H^2(G, H^0(\bar{X}, \bar{F})) \to H^2(X, F). \end{aligned}$$

Another application of the spectral sequence (7.1) in combination with Theorem (6.3.2) is the computation of the étale cohomology of henselian rings.

Let A be a henselian ring (cp. (6.1)) with maximal ideal $\mathfrak{m}$ and residue field k. Let $\bar{k}$ be a separable closure of k and let $G = Gal(\bar{k}/k)$. We take $X = spec(A)$. Then $spec(\bar{k})$ is canonically a geometric point of X. If F is an abelian sheaf on $X_{\acute{e}t}$, then the stalk $F_{spec(\bar{k})}$ has the structure of a continuous G-module.

(7.3) Theorem. *For each abelian sheaf F on $X_{\acute{e}t}$ and for all $q \geq 0$ we have a canonical isomorphism*

$$H^q(X, F) \cong H^q(G, F_{spec(\bar{k})}).$$

Remark. This theorem contains lemma (6.2.3) on the étale cohomology of strictly local rings as a special case. Conversely, using (7.1) and (6.3.2) we can reduce (7.3) to (6.2.3) as follows:

Proof of (7.3): Let T'_G denote the canonical topology on the category of finite continuous left G-sets. We obtain a morphism

$$\varphi : T'_G \to X_{\acute{e}t}$$

of topologies in the following way:

The topology T'_G is equivalent to the restricted étale site of *spec*(k), whose underlying category consists of finitely presented étale *spec*(k)-schemes, hence is the category of affine étale *spec*(k)-schemes. By (6.1.4) this category is equivalent to the category of finite étale X-schemes. Composing both equivalences we obtain the functor $\varphi : T'_G \to X_{\text{ét}}$, which is easily seen to be a morphism of topologies.

Consider now the spectral sequence (7.1) for $\varphi : T'_G \to X_{\text{ét}}$:

$$E_2^{pq} = H^p(G, \varinjlim H^q(X_{\text{ét}}; \varphi(G/H), F)) \Longrightarrow E^{p+q} = H^{p+q}(X, F).$$

We use (6.3.2) to compute the terms $\varinjlim H^q(X_{\text{ét}}; \varphi(G/H), F)$. By construction the schemes $\varphi(G/H)$ are affine. Let $\bar{X} = \varprojlim \varphi(G/H)$, and let $\bar{F}$ denote the inverse image of F under $\bar{X} \to X$. Then we obtain

$$\varinjlim H^q(X_{\text{ét}}; \varphi(G/H), F) \cong H^q(\bar{X}, \bar{F}).$$

Hence our spectral sequence reads:

$$E_2^{pq} = H^p(G, H^q(\bar{X}, \bar{F})) \Longrightarrow E^{p+q} = H^{p+q}(X, F).$$

We claim that $\bar{X} = \varprojlim \varphi(G/H)$ is equal to the strict localization of X in the geometric point *spec*$(\bar{k}) \to X$ of X. Assuming this for the moment, we can apply lemma (6.2.3) and obtain

$$H^q(\bar{X}, \bar{F}) \cong \begin{cases} \bar{F}_{spec(\bar{k})} \cong F_{spec(\bar{k})} & \text{for } q = 0 \\ 0 & \text{for } q > 0. \end{cases}$$

Hence $E_2^{pq} = 0$ for $q > 0$ and therefore the edge morphisms $E_2^{p,0} \to E^p$ are all isomorphisms. But this means

$$H^p(G, F_{spec(\bar{k})}) \to H^p(X, F)$$

is an isomorphism for all $p \geq 0$.

We write $\varphi(G/H) = spec(A(G/H))$, where $A(G/H)$ is the finite étale A-algebra corresponding to the fixed field of the open normal subgroup H under the equivalence (6.1.4). Part ii) of (6.1.1) implies that the A-algebra $A(G/H)$ is local. The construction of the strict henselization A^{hs} of A with respect to $k \to \bar{k}$ (cp. (6.1.9)) now shows that we obtain a canonical homomorphism

$$\varinjlim A(G/H) \to A^{hs}.$$

Assume now that B is an arbitrary essentially étale A-algebra with a k-homomorphism $k(B) \to \bar{k}$. Then $k(B)$ is finite over k, and therefore $k(B) \to \bar{k}$ factors through the inclusion $k' \to \bar{k}$ for some fixed field k' of a suitably chosen open normal subgroup H of G. Since A is henselian, and $B/\mathfrak{m}B = k(B)$ is finite over k, the algebra B is finite over A by (6.1.1) iii). Therefore, by (6.1.4), there is a local A-homomorphism $B \to A(G/H)$ corresponding to the k-homomorphism $k(B) \to k'$, and this shows that the map $\varinjlim A(G/H) \to A^{hs}$ is in fact an isomorphism. □

§8. The Decomposition Theorem Relative Cohomology

(8.1) The Decomposition Theorem

(8.1.1) Proposition. *If $f : X \to Y$ is an* **immersion** *of schemes, then the adjoint morphism (cp. 0, (1.1))*

$$\sigma_G : f^* f_*(G) \to G$$

is an isomorphism for all abelian sheaves G on $Y_{\text{ét}}$. In other words, the functor $f_* : \tilde{Y}_{\text{ét}} \to \tilde{X}_{\text{ét}}$ is fully faithful (see the comments after I, (3.9.2)).

Proof: f is the composite of an open and of a closed immersion. Since the composite of fully faithful functors is again fully faithful, it suffices to deal with each case separately.

Let $f : X \to Y$ be an open immersion. Then in particular f is étale and by (1.4.8) we have for étale Y-schemes Y':

$$f^* f_* G(Y') = f_* G(Y') = G(Y' \times_X Y) \cong G(Y').$$

Let $f : X \to Y$ be a closed immersion. By (5.6) it is enough to show that the maps on the stalks

$$(f^* f_*(G))_{\bar{y}} \to G_{\bar{y}}$$

are bijective for all $y \in Y$. Now by (5.2) iii) we have a canonical isomorphism $(f^*f_*(G))_{\bar{y}} \cong (f_*G)_{\bar{y}}$. Since $f : X \to Y$ is finite as a closed immersion, we can apply Theorem (6.4.2) and obtain $(f_*G)_{\bar{y}} \cong G_{\bar{y}}$. □

In the following let X be a scheme, Y be a closed subscheme of X, and let U be the open complement of Y with the induced structure of a subscheme. Let $i : Y \to X$ and $j : U \to X$ denote the canonical immersions.

We consider the functor $i_* : \widetilde{Y}_{\text{ét}} \to \widetilde{X}_{\text{ét}}$. This is fully faithful by the result just proven, and therefore induces an equivalence between $\widetilde{Y}_{\text{ét}}$ and a full subcategory of $\widetilde{X}_{\text{ét}}$.

Given an abelian sheaf F on $X_{\text{ét}}$, we say that F **vanishes outside** Y if F satisfies the following obviously equivalent conditions:

a) $supp(F) \subseteq Y$
b) $F_{\bar{x}} = 0$ for all $x \in U$
c) $j^*F = 0$
d) $F(X') = 0$ for all étale X-schemes X' satisfying $X' \times_X Y = \emptyset$.

(8.1.2) Theorem. *The functor $i_* : \widetilde{Y}_{\text{ét}} \to \widetilde{X}_{\text{ét}}$ induces an equivalence between $\widetilde{Y}_{\text{ét}}$ and the full subcategory of $\widetilde{X}_{\text{ét}}$ of all abelian sheaves on $\widetilde{X}_{\text{ét}}$, which vanish outside Y.*

Proof: We know already that i_* is fully faithful. It remains to be shown that $F \in \widetilde{X}_{\text{ét}}$ vanishes outside of Y if and only if $F \cong i_*G$ for some $G \in \widetilde{Y}_{\text{ét}}$.

Let $G \in \widetilde{Y}_{\text{ét}}$ be given. Then we have for each étale X-scheme X' with $X' \times_X Y = \emptyset$:

$$i_*G(X') = G(X' \times_X Y) = G(\emptyset) = 0,$$

hence i_*G vanishes outside of Y.

Conversely, let F be an abelian sheaf or $X_{\text{ét}}$, which vanishes outside of Y. Then i^*F is an abelian sheaf on $Y_{\text{ét}}$, and we are done, if we show that the adjoint morphism (cp. 0, (1.1))

$$\rho_F : F \to i_*i^*(F)$$

is an isomorphism. We show this for each stalk:

If $x \in U$, then $F_{\bar{x}} = 0$ and $(i_*i^*F)_{\bar{x}} = 0$, since both F and i_*i^*F vanish outside of Y.

Now take $x \in Y$. By (5.2) iii) it suffices to show that

$$i^*(\rho_F) : i^*F \to i^*(i_*i^*F)$$

is an isomorphism. Now, in general, the composite

$$i^*F \xrightarrow{i^*(\rho_F)} i^*i_*i^*F \xrightarrow{\sigma_{i^*F}} i^*F$$

is the identity morphism of i^*F. Since by (8.1.1) σ_{i^*F} is an isomorphism, the same then holds for $i^*(\rho_F)$. □

(8.1.3) Remark. If the closed subscheme Y has the same underlying topological space as X, in other words: if Y is defined by a locally nilpotent sheaf of ideals of $\mathcal{O}_X$, then we obtain from (8.1.2) that $i_* : \widetilde{Y}_{\text{ét}} \to \widetilde{X}_{\text{ét}}$ is an equivalence of categories. The quasi-inverse functor is then of course given by $i^* : \widetilde{X}_{\text{ét}} \to \widetilde{Y}_{\text{ét}}$. But in fact a more general statement is true (cp. [20], 18.1.2. and [2], exp. VIII, 1.1.): The functor $i_{\text{ét}} : X_{\text{ét}} \to Y_{\text{ét}}$, defined by $X' \mapsto X' \times_X Y$, is an equivalence of topologies.

Remark. Consider the open immersion $j : U \to X$. If G is an abelian sheaf on $U_{\text{ét}}$ and if $x \notin U$, then we may very well have $(j_*G)_{\bar{x}} \neq 0$. If, for instance, X is irreducible and normal, and if U is any non-empty open subset of X, then the canonical morphism of sheaves $\mathbb{Z}_X \to j_*(\mathbb{Z}_U)$ is an isomorphism. Hence $j_*(\mathbb{Z}_U)_{\bar{x}} = \mathbb{Z}$ for all $x \in X$. To see this, observe that with X each étale X-scheme X' is normal as well (cp. [20], 6.5.4.), so that the connected components of X' are irreducible and in a unique correspondence with those of $X' \times_X U$.

Assume we are given an abelian sheaf F on $X_{\text{ét}}$. We then obtain a sheaf $F_U = j^*F$ on $U_{\text{ét}}$, a sheaf $F_Y = i^*F$ on $Y_{\text{ét}}$, and a morphism of sheaves $F_Y \to i^*j_*F_U$, induced by the adjoint morphism $\rho_F : F \to j_*j^*F$. In the following we want to show that first of all the sheaf F is determined by F_U, F_Y and the morphism $F_Y \to i^*j_*F_U$, and that secondly we obtain a sheaf on $X_{\text{ét}}$ from a given sheaf A on $U_{\text{ét}}$, a given sheaf B on $Y_{\text{ét}}$ and a 'pasting' morphism $B \to i^*j^*A$. More precisely we will show: The functor

$$F \mapsto (F_U, F_Y, F_Y \to i^*j_*F_U)$$

is an equivalence between the category $\widetilde{X}_{\text{ét}}$ and the so-called **mapping cylinder of the left exact additive functor** $i^*j_* : \widetilde{U}_{\text{ét}} \to \widetilde{Y}_{\text{ét}}$.

In general, let $f : \mathcal{A} \to \mathcal{B}$ be a left exact additive functor of abelian categories. Following M. Artin, we construct a new category $\mathcal{C}$, the so-called **mapping cylinder of** f, as follows:

a) The objects of $\mathcal{C}$ are triples

$$(A, B, \varphi), \text{ where } \begin{cases} A \in \mathcal{A} \\ B \in \mathcal{B} \\ \varphi : B \to f(A) \quad \text{a morphism in } \mathcal{B}. \end{cases}$$

b) The morphisms $\xi : (A, B, \varphi) \to (A', B', \varphi')$ in $\mathcal{C}$ are defined as pairs $\xi_A : A \to A'$, $\xi_B : B \to B'$, such that the following diagram commutes:

$$\begin{array}{ccc} B & \xrightarrow{\varphi} & f(A) \\ {\scriptstyle \xi_B}\downarrow & & \downarrow{\scriptstyle f(\xi_A)} \\ B' & \xrightarrow{\varphi'} & f(A'). \end{array}$$

It is not hard to show the following:

(8.1.4) The category $\mathcal{C}$ is abelian, and a sequence

$$(A', B', \varphi') \to (A, B, \varphi) \to (A'', B'', \varphi'')$$

is exact in $\mathcal{C}$, if and only if the two sequences

$$A' \to A \to A'' \quad \text{and} \quad B' \to B \to B''$$

are exact.

We define the following additive functors:

$$\mathcal{A} \;\begin{array}{c} \xrightarrow{j_!} \\ \xleftarrow{j^*} \\ \xrightarrow{j_*} \end{array}\; \mathcal{C} \;\begin{array}{c} \xrightarrow{i^*} \\ \xleftarrow{i_*} \\ \xrightarrow{i^!} \end{array}\; \mathcal{B}$$

$$\begin{array}{ll} j^* : \ (A, B, \varphi) \mapsto A; & i^* : (A, B, \varphi) \mapsto B \\ j_* : \ A \mapsto (A, f(A), id); & i_* : B \mapsto (0, B, 0) \\ j_! : \ A \mapsto (A, 0, 0) & \\ i^! : \ (A, B, \varphi) \mapsto ker\ \varphi. & \end{array}$$

We mention the following properties:

(8.1.5) i) Each of the functors in the diagram above is left adjoint to the one below it, e.g.

$$\mathrm{Hom}(j_! X, Y) \cong \mathrm{Hom}(X, j^* Y)$$
$$\mathrm{Hom}(j^* Y, Z) \cong \mathrm{Hom}(Y, j_* Z).$$

ii) The functors $j^*, j_!, i^*, i_*$ are exact, the functors $j_*, i^!$ are left exact.

iii) j_* and i_* are fully faithful.

iv) $i^*j_* = f$ and $i^*j_! = i^!j_! = i^!j_* = j^*j_* = 0$.

(8.1.6) Lemma. *Assume we have additive functors*

$$\mathcal{A} \underset{j_*}{\overset{j^*}{\leftrightarrows}} \mathcal{C} \underset{i_*}{\overset{i^*}{\rightleftarrows}} \mathcal{B}$$

of abelian categories satisfying the following conditions:

a) j^* is left adjoint to j_*, i^* is left adjoint to i_*.

b) j^* and i^* are exact.

c) j_* and i_* are fully faithful.

d) For $C \in \mathcal{C}'$ we have $j^*C = 0$ if and only if $C \cong i_*B$ with $B \in \mathcal{B}$.

Then the functor $f = i^*j_* : \mathcal{A} \to \mathcal{B}$ is left exact and additive, and the functor

$$C \mapsto (j^*C, i^*C, i^*C \xrightarrow{i^*(\rho_C)} i^*j_*j^*C)$$

is an equivalence between the category $\mathcal{C}'$ and the mapping cylinder $\mathcal{C}$ of $f = i^*j_*$.

Proof: Since j_* has a left adjoint functor, it is left exact. By assumption i^* is exact, hence $f = i^*j_* : \mathcal{A} \to \mathcal{B}$ is left exact.

Let $F : \mathcal{C}' \to \mathcal{C}$ denote the functor from the category $\mathcal{C}'$ to the mapping cylinder $\mathcal{C}$ of $f = i^*j_* : \mathcal{A} \to \mathcal{B}$ defined in the statement of the lemma.

We define a functor $G : \mathcal{C} \to \mathcal{C}'$ by

$$G : (A, B, \varphi) \to j_*A \times_{i_*i^*j_*A} i_*B,$$

where the fibre product is taken with respect to the morphisms $\rho_{j_*A} : j_*A \to i_*i^*j_*A$ and $i_*(\varphi) : i_*B \to i_*i^*j_*A$.

We claim that the functor $G : \mathcal{C} \to \mathcal{C}'$ is quasi-inverse to $F : \mathcal{C}' \to \mathcal{C}$, hence that $G \circ F = id_{\mathcal{C}'}$ and $F \circ G = id_{\mathcal{C}}$. Let us first prove the following statement:

A morphism $\varepsilon : C \to C'$ in $\mathcal{C}'$ is an isomorphism if and only if both $j^*\varepsilon$ and $i^*\varepsilon$ are isomorphisms.

Since j^* and i^* are exact by assumption, it suffices to show that for an object $C \in \mathcal{C}'$ we have $C = 0$ if and only if $j^*C = 0$ and $i^*C = 0$. Assume then that $j^*C = 0 = i^*C$. By condition d) there is $B \in \mathcal{B}$, s.t. $C \cong i_*B$. Since the functor $i_* : \mathcal{B} \to \mathcal{C}'$ is fully faithful by assumption c), the adjoint morphism $\sigma_B : i^*i_*B \to B$ is an isomorphism. Hence $B \cong i^*C = 0$, and therefore $C \cong i_*B = 0$.

To prove that $G \circ F = id_{\mathcal{C}'}$, we show that the following commutative diagram of canonical morphisms, which are functorial in C, is cartesian:

$$\begin{array}{ccc} C & \xrightarrow{\rho} & j_*j^*C \\ {\scriptstyle\rho}\downarrow & & \downarrow{\scriptstyle\rho} \\ i_*i^*C & \xrightarrow{\rho} & i_*i^*j_*j^*C \end{array}$$

By what we just proved, this is equivalent to showing that the diagrams obtained by applying j^* and i^* are cartesian.

Let us first apply j^*. Then both objects in the bottom row are 0 by assumption d). Moreover, the composite

$$j^*C \xrightarrow{j^*(\rho_C)} j^*j_*j^*C \xrightarrow{\sigma_{j^*C}} j^*C$$

is the identity of j^*C. Since by assumption c) j_* is fully faithful, and therefore σ_{j^*C} an isomorphism, the same is true for $j^*(\rho_C)$. Hence the diagram is cartesian after applying j^*.

If we apply i^*, a similar argument, using the fact that i_* is fully faithful, shows that the vertical arrows become isomorphisms, and hence the diagram is again cartesian.

The simple proof that $F \circ G = id_{\mathcal{C}}$ is left to the reader. □

Let us return now to the situation considered at the beginning of this section: X is a scheme, Y is a closed subscheme of X, and U is the open complement of Y in X endowed with the canonical structure of a subscheme. Moreover, $i : Y \to X$ and $j : U \to X$ are the canonical immersions.

(8.1.7) Theorem *(The Decomposition Theorem).*
The functor

$$F \mapsto (j^*F, i^*F, i^*F \xrightarrow{i^*(\rho_F)} i^*j_*j^*F)$$

is an equivalence between the category $\widetilde{X}_{\text{ét}}$ *and the mapping cylinder of the functor* $i^*j_* : \widetilde{U}_{\text{ét}} \to \widetilde{Y}_{\text{ét}}$. *The latter is the category of triples*

$$(A, B, \varphi) \quad \text{with} \begin{cases} A \in \widetilde{U}_{\text{ét}} \\ B \in \widetilde{Y}_{\text{ét}} \\ \varphi : B \to i^*j_*A \quad \text{a morphism.} \end{cases}$$

Proof: We have the following functors of abelian categories

$$\widetilde{U}_{\text{ét}} \underset{j_*}{\overset{j^*}{\leftrightarrows}} \widetilde{X}_{\text{ét}} \underset{i_*}{\overset{i^*}{\rightleftarrows}} \widetilde{Y}_{\text{ét}}.$$

These satisfy the conditions a) – d) in Proposition (8.1.6) in view of (1.4.2) i), ii), (8.1.1) and (8.1.2). □

(8.1.8) Example. Under the equivalence of the Decomposition Theorem the constant sheaf $\mathbb{Z}_X$ on $\widetilde{X}_{\text{ét}}$ corresponds to the triple

$$(\mathbb{Z}_U, \mathbb{Z}_Y, \mathbb{Z}_Y \to i^*j_*\mathbb{Z}_U),$$

where the morphism $\mathbb{Z}_Y \to i^*j_*\mathbb{Z}_U$ is induced by the adjoint morphism $\mathbb{Z}_X \to j_*j^*\mathbb{Z}_X$. In particular, if X is normal and irreducible and $U \neq \emptyset$, the morphism $\mathbb{Z}_Y \to i^*j_*\mathbb{Z}_U$ is an isomorphism. This is not true in general. In fact, there is an exact sequence

$$0 \to i^!\mathbb{Z}_X \to \mathbb{Z}_Y \to i^*j_*\mathbb{Z}_U \to R^1i^!(\mathbb{Z}_X) \to 0$$

(cp. the next section (8.2)).

(8.1.9) Example. Let A be a discrete valuation ring with a perfect residue field k. We want to determine the category $\widetilde{X}_{\text{ét}}$ of abelian étale sheaves on the scheme $X = spec(A)$.

Let K be the quotient field of A, $\overline{K}$ be a separable closure of K, and let G denote the Galois group of $\overline{K}/K$. Moreover, let $\bar{v}$ be an extension of the discrete valuation on K to $\overline{K}$, and let $D(\bar{v})$ and $I(\bar{v})$ denote the inertia group and the decomposition group of $\bar{v}$ respectively. By definition these groups are:

$$D(\bar{v}) = \{\sigma \in G \mid \sigma\bar{v} = \bar{v}\}$$
$$I(\bar{v}) = \{\sigma \in D(\bar{v}) \mid \bar{v}(\sigma x - x) > 0 \text{ for all } \bar{v}\text{-integral } x \in \overline{K}\}.$$

As is well known, the residue field $\bar{k}$ of the valuation $\bar{v}$ is an algebraic closure of k, and the canonical homomorphism

$$D(\bar{v})/I(\bar{v}) \to g = Gal(\bar{k}/k)$$

is an isomorphism of topological groups. Therefore, the assignment

$$M \mapsto M^{I(\bar{v})}$$

defines an additive functor from the category of continuous G-modules to the category of continuous g-modules. We have the following result:

The category $\widetilde{X}_{\text{ét}}$ of abelian étale sheaves on $X = spec(A)$ is equivalent to the mapping cylinder of the functor $M \mapsto M^{I(\bar{v})}$, hence to the category of triples

$$(M, N, \varphi), \quad \text{where} \begin{cases} M & \text{is a continuous } G\text{-module} \\ N & \text{is a continuous } g\text{-module} \\ \varphi : N \to M^{I(\bar{v})} & \text{is a } g\text{-homomorphism.} \end{cases}$$

This description of $\widetilde{X}_{\text{ét}}$ follows from the Decomposition Theorem (8.1.7), applied to the closed immersion $i : spec(k) \to X$ and the open immersion $j : spec(K) \to X$. By (2.2) the categories of abelian sheaves on $spec(K)_{\text{ét}}$ and $spec(k)_{\text{ét}}$ are equivalent to the categories of continuous G-modules and g-modules respectively. It is easy to verify that under this equivalence the functor $i^* j_*$ gets transformed into the functor $M \mapsto M^{I(\bar{v})}$.

As an exercise the reader should determine the triples corresponding to the sheaves $(\mathbb{G}_a)_X$, $(\mathbb{G}_m)_X$, $(\mu_n)_X$.

(8.2) The functors $j_!$ and $i^!$

We keep the notations of the previous section: X is a scheme, Y is a closed subscheme of X, U is the open complement of Y in X, $i : Y \to X$ and $j : U \to X$ are the canonical immersions.

The Decomposition Theorem (8.1.7) implies that we have the two functors

$$j_! : \widetilde{U}_{\text{ét}} \to \widetilde{X}_{\text{ét}}$$
$$i^! : \widetilde{X}_{\text{ét}} \to \widetilde{Y}_{\text{ét}},$$

defined by $j_! : A \mapsto (A, 0, 0)$ and $i^! : (A, B, \varphi) \mapsto ker\ \varphi$.

The functor $j_!$ is exact and left adjoint to j^*, the functor $i^!$ is left exact and has i_* as left adjoint functor (cp. (8.1.5)).

(8.2.1) Proposition. *The sequences*

$$0 \to j_!j^*F \to F \to i_*i^*F \to 0$$
$$0 \to i_*i^!F \to F \to j_*j^*F$$

are exact for all abelian sheaves F *on* $X_{\text{ét}}$. *Here the non-trivial arrows are the adjoint morphisms.*

Proof: The equivalence of categories (8.1.7) transforms the sequences into

$$0 \to (A, 0, 0) \to (A, B, \varphi) \to (0, B, 0) \to 0$$
$$0 \to (0, ker\ \varphi, 0) \to (A, B, \varphi) \to (A, i^*j_*A, id)$$

Both these sequences are exact (cp. (8.1.4)). □

Remark. We will determine the cokernel of the adjoint morphism $F \to j_*j^*F$ in (8.2.4).

By definition we have $j^*j_!G = G$ and $i^*j_!G = 0$ for an abelian sheaf G on $U_{\text{ét}}$. Hence by (5.2) iii) the stalks $(j_!G)_{\bar{x}}$ for points $x \in X$ are computed as follows:

$$(j_!G)_{\bar{x}} = \begin{cases} G_{\bar{x}} & \text{for } x \in U \\ 0 & \text{for } x \in X \setminus U. \end{cases}$$

The sheaf $j_!G$ is called the **extension of** G **by** 0.

For an abelian sheaf F on $X_{\text{ét}}$ the sheaf $i_*i^!F$ gets identified by (8.2.1) with the subsheaf $ker(F \to j_*j^*F)$ of F, which vanishes outside Y. The sheaf $i_*i^!F$ is maximal with this property: If G is another subsheaf of F, which vanishes outside Y, then we have $j^*G = 0$, hence $G \to j_*j^*G$ is the zero map. If we identify $\widetilde{Y}_{\text{ét}}$ by means of i_* with the full subcategory of $\widetilde{X}_{\text{ét}}$ of all abelian sheaves vanishing outside Y (cp. (8.1.2)), then we see that $i^! F$ is the **maximal subsheaf of F, which vanishes outside Y.** Another terminology is the following: $i^! F$ is the "subsheaf of all sections of F with support in Y".

As we mentioned already, the functor $i^! : \widetilde{X}_{\text{ét}} \to \widetilde{Y}_{\text{ét}}$ is left exact. In order to determine the right derived functors $R^q i^!$, we need the following:

(8.2.2) Lemma. *For a* **flabby** *abelian sheaf F on $X_{\text{ét}}$ the sequence*

$$0 \to i_* i^! F \to F \to j_* j^* F \to 0$$

is exact in the category of **abelian presheaves** *on $X_{\text{ét}}$ (hence a fortiori in the category of abelian sheaves on $X_{\text{ét}}$).*

Proof: In view of (8.2.1) it only remains to be shown that the map $F(X') \to j_* j^* F(X')$ is surjective for all étale X-schemes X'. If we define the open subset U' of X' by $U' = U \times_X X'$, then we obtain $j_* j^* F(X') = j_* F(U') = F(U')$, where the last equality follows from (1.4.8). Therefore the adjoint morphism $F(X') \to j_* j^* F(X')$ is identified with the restriction map $F(X') \to F(U')$ induced by $U' \subset X'$. Hence the following more general statement, which is also interesting in itself, will complete the proof of (8.2.2).

(8.2.3) Proposition. *Let F be a flabby abelian sheaf on $X_{\text{ét}}$, let X' be an étale X-scheme, and let $U' \subset X'$ be open. Then the restriction map $F(X') \to F(U')$ is surjective.*

Proof: Without loss of generality (cp. (1.4.8)) we may assume that $X' = X$. Assume $U \subset X$ is open, but $F(X) \to F(U)$ is not surjective. Let Z denote the scheme obtained by pasting two copies X_1 and X_2 of X together along U. The scheme Z is étale over X, $\{X_i \to Z\}_{i=1,2}$ is a covering in $X_{\text{ét}}$, and $X_1 \times_Z X_2 = U$. For $t \in F(U)$, but $t \notin im(F(X) \to F(U))$, consider the 1-cochain $\mathrm{t} \in C^1(\{X_i \to Z\}, F)$ with components $0, t, -t, 0$ in $F(X_1 \times_Z X_1)$, $F(X_1 \times_Z X_2)$, $F(X_2 \times_Z X_1)$ and $F(X_2 \times_Z X_2)$. It is easy to see that t is a cocycle, but not a coboundary. Therefore $H^1(\{X_i \to Z\}, F) \neq 0$, contradicting the fact that F is flabby. □

(8.2.4) Corollary. *i) For all abelian sheaves F on $X_{\text{ét}}$ the sequence*

$$0 \to i_* i^! F \to F \to j_* j^* F \to i_* R^1 i^!(F) \to 0$$

is exact.

ii) For all abelian sheaves F on $X_{\text{ét}}$:

$$i_* R^q i^!(F) \cong R^{q-1} j_*(j^* F) \quad \text{for } q \geq 2.$$

Proof: We choose an injective resolution $M^\cdot = (M^i)_{i=0,1,\dots}$ of F:

$$0 \to F \to M^0 \to M^1 \to \cdots$$

Since each M^i is flabby as an injective sheaf, we obtain from (8.2.2) the following exact sequence of complexes:

$$0 \to i_* i^!(M^\cdot) \to M^\cdot \to j_* j^*(M^\cdot) \to 0.$$

The corresponding long exact sequence is

$$\cdots \to R^{q-1} id(F) \to R^{q-1}(j_* j^*)(F) \to R^q(i_* i^!)(F) \to R^q id(F) \to \cdots$$

But $R^q id(F) = 0$ for $q > 0$, and $R^q(j_* j^*)(F) = R^q j_*(j^* F)$ as well as $R^q(i_* i^!)(F) = i_* R^q i^!(F)$, since both functors j^* and i_* are exact. Therefore, for $q = 1$ we obtain the exact sequence i), and for $q \geq 2$ the isomorphism ii). □

(8.3) Relative Cohomology

As before, let X be a scheme, Y be a closed subscheme with open complement U in X, and let $i : Y \to X$ and $j : U \to X$ denote the canonical immersions.

For an abelian sheaf F on $X_{\text{ét}}$ we write:

$$H^0_Y(X, F) = H^0(X, i_* i^! F) = H^0(Y, i^! F).$$

We also have $H^0_Y(X, F) = ker(F(X) \to F(U))$ by (8.2.1). This means that $H^0_Y(X, F)$ consists of all sections s in $F(X)$, such that $supp(s) \subset Y$.

As the composite of the left exact functors $i^!$ and $H^0(Y, \cdot)$, the functor $F \mapsto H^0_Y(X, F)$ is left exact. We define

$$H^q_Y(X, \cdot) := R^q H^0_Y(X, \cdot).$$

The groups $H^q_Y(X,F)$ are called the **relative** (or **local**) **cohomology groups of F with support in Y.**

We have to distinguish between the relative cohomology groups $H^q_Y(X,F)$ and the cohomology groups $H^q(X,i_*i^!F) = H^q(Y,i^!F)$. The relation between them is given by a spectral sequence: Since the left adjoint functor i_* of $i^!$ is exact, the functor $i^!$ maps injective objects of $\widetilde{X}_{\text{ét}}$ to injective objects of $\widetilde{Y}_{\text{ét}}$, and therefore we obtain (cp. 0, (2.3.5)) the spectral sequence

$$H^p(Y,R^qi^!F) \Longrightarrow H^{p+q}_Y(X,F). \tag{8.3.1}$$

In general, $R^qi^!(F) \neq 0$ for $q > 0$ and the edge morphisms $H^p(Y,i^!F) \to H^p_Y(X,F)$ are not isomorphisms.

(8.3.2) Theorem *(Relative cohomology sequence).*
For all abelian sheaves F on $X_{\text{ét}}$ there is a long exact sequence

$$\cdots \to H^q_Y(X,F) \to H^q(X,F) \to H^q(U,F) \to H^{q+1}_Y(X,F) \to \cdots$$

Proof: By (8.2.2) the sequence

$$0 \to i_*i^!F \to F \to j_*j^*F \to 0$$

is exact in the category of abelian presheaves on $X_{\text{ét}}$ for all injective abelian sheaves F on $X_{\text{ét}}$. Therefore the sequence

$$0 \to H^0_Y(X,F) \to H^0(X,F) \to H^0(U,F) \to 0$$

is exact.

Now, given an injective resolution $M^\cdot = (M^i)_{i=0,1,\dots}$ of an arbitrary abelian sheaf F on $X_{\text{ét}}$, we obtain an exact sequence of complexes

$$0 \to H^0_Y(X,M^\cdot) \to H^0(X,M^\cdot) \to H^0(U,M^\cdot) \to 0,$$

and the corresponding long exact sequence is the relative cohomology sequence of the theorem. □

§9. Torsion Sheaves, Locally Constant Sheaves, Constructible Sheaves

(9.1) Torsion Sheaves

An abelian sheaf F on an arbitrary topology T is called a **torsion sheaf** if it satisfies the following equivalent conditions:

i) F is a sheaf associated to a presheaf of abelian torsion groups.

ii) The canonical morphism $\varinjlim {}_nF \to F$ is an isomorphism, where ${}_nF$ denotes the kernel of multiplication $F \xrightarrow{n} F$ for $n \in \mathbb{N}$.

The equivalence of i) and ii) can be seen as follows: Let $F = P^{\#}$ with P a presheaf of abelian torsion groups on T. Then the sequence $0 \to {}_nP \to P \xrightarrow{n} P$ is exact in the category of abelian presheaves, and therefore (cp. I, (3.2.1)) the sequence $0 \to ({}_nP)^{\#} \to F \xrightarrow{n} F$ is exact in the category of abelian sheaves. Hence ${}_nF = ({}_nP)^{\#}$. In the category of abelian presheaves we have $P = \varinjlim {}_nP$. Since $\#$ commutes with inductive limits as a left adjoint functor (cp. 0, (3.1.3)), we obtain

$$F = P^{\#} = (\varinjlim {}_nP)^{\#} = \varinjlim {}_nF.$$

Conversely, let us assume that $\varinjlim {}_nF = F$. By I, (3.2.3) the sheaf $\varinjlim {}_nF$ is associated to the presheaf $U \mapsto \varinjlim {}_nF(U)$ on T, and $\varinjlim {}_nF(U)$ is a torsion group, being the inductive limit of the torsion groups ${}_nF(U)$.

If F is a torsion sheaf on T, the groups $F(U)$ for $U \in T$ need not be torsion groups. This is however the case, if U is **quasi-compact**. Recall (cp. I, (3.10.1)) that this means that each covering of U has a finite subcovering. In fact, if P denotes the presheaf limit $U' \mapsto \varinjlim {}_nF(U')$, then P is separated as a subpresheaf of F. Therefore I, (3.1.3) ii) and the quasi-compactness of U yield

$$\begin{aligned} F(U) = P^{\dagger}(U) = \check{H}^0(U,P) &= \varinjlim_{\substack{\{U_i \to U\} \\ \text{finite}}} H^0(\{U_i \to U\}, P) \\ &= \varinjlim_{\substack{\{U_i \to U\} \\ \text{finite}}} ker(\prod_i P(U_i) \rightrightarrows \prod_{i,j} P(U_i \times_U U_j)). \end{aligned}$$

This shows that $F(U)$ is a torsion group.

Let X be a scheme. Then we have:

(9.1.1) Proposition. *An abelian sheaf F on $X_{\text{ét}}$ is a torsion sheaf if and only if all stalks $F_{\bar{x}}$ are torsion groups.*

Proof: Since the stalk functors are exact (cp. (5.2) i)), the group $({}_nF)_{\bar{x}}$ is equal to the kernel ${}_n(F_{\bar{x}})$ of $F_{\bar{x}} \xrightarrow{n} F_{\bar{x}}$. Moreover, since the stalk functors commute with inductive limits (cp. (5.2), i)), it follows from (5.6), i) that $\varinjlim {}_nF \to F$ is an isomorphism if and only if $\varinjlim {}_n(F_{\bar{x}}) \to F_{\bar{x}}$ is bijective for all $x \in X$, hence if and only if $F_{\bar{x}}$ is a torsion group for all $x \in X$. □

If the scheme X is quasi-compact (and hence quasi-compact in the meaning of I, (3.10,1) as an object of $X_{\text{ét}}$), and if F is a torsion sheaf on $X_{\text{ét}}$, then the group $H^0(X, F)$ is torsion. Furthermore we have

(9.1.2) Proposition. *If the scheme X is* **quasi-compact** *and* **quasi-separated**, *and if F is a torsion sheaf on $X_{\text{ét}}$, then the cohomology groups $H^q(X, F)$ are torsion for all q.*

Proof: We have $F = \varinjlim {}_nF$. Since by assumption X is quasi-compact and quasi-separated, (1.5.3) implies that $H^q(X, F) = \varinjlim H^q(X, {}_nF)$. Therefore it suffices to prove the statement for $F = {}_nF$. In this case the map $F \xrightarrow{n} F$ zero map, and hence the same is true for the induced map $H^q(X, F) \to H^q(X, F)$. But this induced map in cohomology is also given by multiplication by n, hence $H^q(X, F)$ is a torsion group. □

(9.1.3) Proposition. *i) If $f : X \to Y$ is a morphism of schemes and if F is a torsion sheaf on $Y_{\text{ét}}$, then f^*F is a torsion sheaf on $X_{\text{ét}}$.*

ii) If $f : X \to Y$ is quasi-compact and quasi-separated and if F is a torsion sheaf on $X_{\text{ét}}$, then $R^q f_* F$ is a torsion sheaf on $Y_{\text{ét}}$ for all q.

Proof: i) Let $x \in X$ and let $f(x) = y$. Then we have $(f^*F)_{\bar{x}} = F_{\bar{y}}$ (cp. (5.2)), and hence the claim follows from (9.1.1).

ii) Since by assumption $f : X \to Y$ is quasi-compact and quasi-separated, we have for all $y \in Y$:

$$(R^q f_*(F))_{\bar{y}} \cong H^q(\bar{X}, \bar{F})$$

by (6.4.1). Here $\overline{X} = X \times_Y \overline{Y}$, where $\overline{Y}$ is the strict localization of Y in $\bar{y}$, and $\overline{F}$ denotes the inverse image of F under $\overline{X} \to X$. By part i) the sheaf $\overline{F}$ is a torsion sheaf on $\overline{X}$. Since $\overline{X}$ is quasi-compact and quasi-separated as a consequence of $\overline{X} \to \overline{Y}$ being quasi-compact and quasi-separated and $\overline{Y}$ being affine, the groups $H^q(\overline{X}, \overline{F})$ are torsion by (9.1.2). Again, the claim follows from (9.1.1). □

Similar considerations of stalks also yield the following:

(9.1.4) Addendum. i) If $j : U \to X$ is an open immersion, then the functor $j_!$ (cp. (8.2)) maps torsion sheaves to torsion sheaves.

ii) If $i : Y \to X$ is a closed immersion, then the functor $i^!$ (cp. loc. cit.) maps torsion sheaves to torsion sheaves. The same property holds for the functors $R^q i^!$ for $q \geq 1$ if the open set $X \smallsetminus Y$ is retro-compact (cp. the definition in (9.3)) or – equivalently – if Y is constructible in X (cp. (9.3) and [19] 0_{III}, 9.1.1. – 9.1.5.). This follows from (8.2.4).

Examples of torsion sheaves on $X_{\text{ét}}$ are given by the sheaves $(\mu_n)_X$ of n-th roots of unity and by the constant sheaves A_X defined by a discrete abelian torsion group A. More examples can be obtained as follows:

Let X be a noetherian scheme, let x be a point in X and let $i : spec(k(x)) \to X$ be the canonical morphism. Given an arbitrary abelian sheaf F on $spec(k(x))_{\text{ét}}$, the sheaves $R^q i_*(F)$ are torsion sheaves on $X_{\text{ét}}$ for $q \geq 1$. To see this we consider the stalks $(R^q i_*(F))_{\bar{y}}$ for points $y \in X$. The morphism $i : spec(k(x)) \to X$ is quasi-compact and quasi-separated, and from (6.4.1) we obtain:

$$(R^q i_*(F))_{\bar{y}} \cong H^q(X(\bar{y}) \times_X spec(k(x)), \overline{F}).$$

Here $X(\bar{y})$ is the strict localization of X in $\bar{y}$ and $\overline{F}$ is the inverse image of F under $X(\bar{y}) \times_X spec(k(x)) \to spec(k(x))$. The scheme $X(\bar{y}) \times_X spec(k(x))$ is the fibre of $X(\bar{y}) \to X$ in the point x. Since X is noetherian by assumption, the fibre of $X(\bar{y}) \to X$ in x is equal to a finite sum of spectra of separable algebraic extensions of $k(x)$ (cp. [20], 18.8.12., ii)). By Galois cohomology, the cohomology groups over the spectrum of a field are torsion groups in dimensions ≥ 1 (cp. [32], I. 2.2.). More generally, this also holds for a finite sum of spectra of fields. Therefore the groups $H^q(X(\bar{y}) \times_X spec(k(x)), \overline{F})$ are torsion for $q \geq 1$, and the result on the sheaves $R^q i_*(F)$ follows from 9.1.1.

Using this result we now can show the following property for the sheaf i_*F, which need not be a torsion sheaf:

For $p > 0$ the cohomology groups $H^p(X, i_*F)$ are torsion.

To see this, we consider the Leray spectral sequence (1.4.3)

$$H^p(X, R^q i_*(F)) \Longrightarrow H^{p+q}(spec(k(x)), F).$$

Being noetherian, the scheme X is quasi-compact and quasi-separated (cp. [17], 6.1.13.), and hence by (9.1.2) the groups $H^p(X, R^q i_*(F))$ are torsion for $p \geq 0$ and $q > 0$. As noted above, the groups $H^n(spec(k(x)), F)$ are torsion as well for $n > 0$. Now, if in general $E_2^{pq} \Longrightarrow E^{p+q}$ is a cohomological spectral sequence, where the initial terms E_2^{pq} are torsion groups for $p \geq 0$ and $q > 0$ and the limit terms E^n are torsion groups for $n > 0$, the terms $E_2^{p,0}$ are torsion as well for $p > 0$: First of all we have a monomorphism $E_\infty^{p,0} \to E^p$ (cp. the construction of the edge morphisms in 0, (2.3.1)), hence for $p > 0$ the groups $E_\infty^{p,0}$ are torsion, and therefore so are the groups $E_r^{p,0}$ for r sufficiently large. The defining property c) for spectral sequences (cp. 0, (2.3)) implies that the torsion property of the terms $E_2^{p,q}$ for $p \geq 0$, $q > 0$ carries over to the terms E_r^{pq} for all $r \geq 2$. The same property c) implies then finally, that for $r \geq 2$ the torsion property of $E_{r+1}^{p,0}$ carries over to the term $E_r^{p,0}$.

As an application we want to show:

(9.1.5) Theorem. *For a regular noetherian scheme X the cohomology groups $H^q(X, (\mathbb{G}_m)_X)$ are torsion for $q \geq 2$.*

Proof: We use the exact sequence

$$0 \to (\mathbb{G}_m)_X \to j_*(\mathbb{G}_m)_K \to \mathcal{D}iv_X \to 0,$$

where K denotes the ring of rational functions on X and $j : spec(K) \to X$ denotes the canonical morphism (cp. (4.5)). We saw above that the groups

$H^q(X, j_*(\mathbb{G}_m)_K)$ are torsion for $q > 0$. Since X is regular, we have (cp. loc. cit.)

$$\mathcal{D}iv_X \cong \bigoplus_{x \in X^{(1)}} (i_X)_* \mathbb{Z}_X,$$

where $X^{(1)}$ is the set of all points $x \in X$ with $\dim \mathcal{O}_{X,x} = 1$, $i_X : spec(k(x)) \to X$ is the canonical morphism, and $\mathbb{Z}_X$ is the constant sheaf on $spec(k(x))_{\text{ét}}$ with value $\mathbb{Z}$. Since $H^q(X, \cdot\,)$ commutes with direct sums (cp. (1.5.3)), we conclude that the cohomology groups $H^q(X, \mathcal{D}iv_X)$ are torsion as well for $q > 0$. The claim follows from the long exact cohomology sequence attached to the exact sequence above. □

(9.1.6) Remark. As is well known, an abelian group is called a **ℓ-torsion group** for a prime number ℓ if each element is annihilated by a power of ℓ. In an obvious manner the notion of a **ℓ-torsion sheaf** on a topology can be defined, and the results (9.1.1) – (9.1.4) have analogs for ℓ-torsion sheaves.

Let X be a scheme, and let F be a torsion sheaf on $X_{\text{ét}}$. Given a prime number ℓ, we look at the subsheaf $F(\ell) = \varinjlim {}_{\ell^n}F$, where ${}_{\ell^n}F$ denotes the kernel of multiplication by ℓ^n. The sheaf $F(\ell)$ is a ℓ-torsion sheaf, associated to the presheaf

$$X' \to \varinjlim ker(F(X') \xrightarrow{\ell^n} F(X')).$$

The canonical morphism of sheaves $\bigoplus_\ell F(\ell) \to F$, ℓ running through all prime numbers, is an isomorphism:

$$\bigoplus_\ell F(\ell) \cong F.$$

To see this, we observe that for a geometric point $\bar{x}$ on X the stalk $F(\ell)_{\bar{x}}$ is isomorphic by (5.2), i) to the ℓ-primary component of $F_{\bar{x}}$. Therefore, $\bigoplus F(\ell)_{\bar{x}}$ is the canonical decomposition of the torsion group $F_{\bar{x}}$ (cp. (9.1.1)) into ℓ-primary components.

If we assume now that X is quasi-compact and quasi-separated, (1.5.3) implies that the above isomorphism induces an isomorphism

$$H^q(X, F) \cong \bigoplus_\ell H^q(X, F(\ell)).$$

Therefore by (9.1.2) the group $H^q(X, F(\ell))$ is isomorphic to the ℓ-primary component of $H^q(X, F)$.

(9.1.7) Definition. *Let X be a quasi-compact and quasi-separated scheme. Let ℓ be a prime number. The* **cohomological ℓ-dimension of X** *is the smallest natural number $cd_\ell(X) = n$ (or ∞, if it does not exist), such that the following conditions hold:*

i) $H^q(X, F) = 0$ for all $q > n$ and for all ℓ-torsion sheaves F on $X_{ét}$.

ii) $H^q(X, F)(\ell) = 0$ for all $q > n$ and for all torsion sheaves F on $X_{ét}$.

(Here we use the notation $A(\ell)$ for the ℓ-primary component of an abelian group A.)

Furthermore

$$cd(X) = \max_\ell(cd_\ell(X))$$

is called the **cohomological dimension of X**. *It is the smallest natural number (or ∞), such that $H^q(X, F) = 0$ for all $q > n$ and all torsion sheaves F on $X_{ét}$.*

In case that $X = spec(A)$ we also write $cd_\ell(A)$ and $cd(A)$ instead of $cd_\ell(X)$ and $cd(X)$.

(9.2) Locally Constant Sheaves

(9.2.1) Definition. *An abelian sheaf F on $X_{ét}$ is called* **locally constant** *if there is a covering $\{X_i \to X\}$ of X in $X_{ét}$ such that the restrictions F/X_i (cp. (1.4.8)) are constant.*

It is immediately clear that inverse images of locally constant sheaves under morphisms of schemes are again locally constant. However, for direct images this is not necessarily true: If for instance X is the scheme of a local ring, x is the closed point of X, $i : spec(k(x)) \to X$ is the canonical immersion and $\mathbb{Z}_x$ is the constant sheaf on $spec(k(x))_{ét}$ with value $\mathbb{Z}$, the sheaf $i_*(\mathbb{Z}_x)$ is not locally constant on $X_{ét}$. The reason is that $i_*(\mathbb{Z}_x)$ is concentrated in the closed point x, whereas a locally constant sheaf on $X_{ét}$, which is non-trivial in x, is non-trivial in all stalks.

In the special case $X = spec(k)$, k a field, we can easily describe locally constant abelian sheaves: Let $\bar{k}$ be a separable closure of k, and let G be the Galois group of $\bar{k}/k$. If A is an abelian group, the abelian sheaves on $spec(k)_{\acute{e}t}$, which have A as the stalk in the geometric point $spec(\bar{k})$, correspond by (2.2) to the different G-module structures on A. The constant sheaf with stalk A corresponds to the trivial G-module structure of A. The locally constant sheaves with stalk A correspond uniquely to those continuous G-module structures of A, which factor through a finite quotient of G.

Now let X be an arbitrary scheme. As we observed earlier (cp. (3.1.4)), the sheaf $(\mu_n)_X$ of n-th roots of unity is locally constant if n is invertible on X. More precisely, this sheaf is isomorphic to $(\mathbb{Z}/n\mathbb{Z})_X$. In general we show:

(9.2.2) Proposition. *If G is a commutative, finite and étale group scheme on X, the sheaf G_X represented by G is locally finite on $X_{\acute{e}t}$.*

Proof: We show: If G is a finite and étale X-scheme, then there is a covering $\{X_i \to X\}$ in $X_{\acute{e}t}$, such that $G \times_X X_i \to X_i$ is trivial for each i, i.e. X_i-isomorphic to $\amalg X_i$. It suffices to prove the following: Given a finite étale A-algebra B of constant rank n, there is an étale A-algebra C, such that $B \otimes_A C \cong C \times \cdots \times C$.

We proceed by induction on the rank n. If $n = 1$ there is nothing to prove. We assume now that the statement is correct for all finite étale A-algebras of rank $< n$.

We form $B \otimes_A B$. B becomes a $B \otimes_A B$-algebra via $\delta : b \otimes b' \to bb'$. Since B/A is finite and étale, B is a projective $B \otimes_A B$-module (cp. [20], 18.3.1.). Therefore the epimorphism $\delta : B \otimes_A B \to B$ has a section $s : B \to' B \otimes_A B$ (of $B \otimes_A B$-modules). Let $s(1) = e$. The element e is a non-zero idempotent in $B \otimes_A B$. To see this, let I be the kernel of $\delta : B \otimes_A B \to B$. I is generated by elements of the form $(b \otimes 1 - 1 \otimes b)$ and

$$\begin{aligned}(b \otimes 1 - 1 \otimes b)e &= (b \otimes 1 - 1 \otimes b)s(1) \\ &= s(b) - s(b) = 0.\end{aligned}$$

Hence $I \cdot e = 0$. Since $\delta(1-e) = \delta(1) - \delta(s(1)) = 0$, we have $(1-e) \in I$, hence $(1-e)e = 0$. The idempotent e yields a decomposition of $B \otimes_A B$ into a direct sum of B-algebras:

$$B \otimes_A B = (B \otimes_A B)e \bigoplus (B \otimes_A B)(1-e).$$

Note that $(B \otimes_A B)e \cong B$. Since $B \otimes_A B$ is finite and étale as a B-algebra, the same is true for the B-algebra $(B \otimes_A B)(1-e)$, which has rank $n-1$. By assumption there is an étale B-algebra C, such that

$$(B \otimes_A B)(1-e) \otimes_B C \cong C \times \cdots \times C.$$

C is étale over A as well, and we obtain

$$\begin{aligned} B \otimes_A C &= (B \otimes_A B) \otimes_B C \\ &\cong (B \oplus (B \otimes_A B)(1-e)) \otimes_B C \\ &\cong C \times \cdots \times C. \end{aligned}$$

□

The abelian sheaf G_X on $X_{\text{ét}}$, represented by a commutative, finite, étale group scheme G over X, has finite stalks everywhere. More precisely, the stalk $G_{X,\bar{x}}$ equals the group of $\overline{k(x)}$-rational points of G. Conversely, a locally constant abelian sheaf with finite stalks is of the form considered in (9.2.2):

(9.2.3) Proposition. *Each locally constant abelian sheaf F on $X_{\text{ét}}$ is represented by a (unique) commutative, étale group scheme G over X. If in addition F has finite stalks, then the group scheme G is finite over X.*

Proof (sketch): There exists a covering $\{X_i \to X\}$ in $X_{\text{ét}}$, such that the restrictions F/X_i are constant for all i. F/X_i is represented by a trivial commutative, étale X_i-group scheme G_i. The family of these X_i-schemes G_i carries a canonical descent datum with respect to the morphisms $X_i \to X$ (cp. [7], exp. IV, 2.). Using [7], exp. X, 5.4, we conclude that this descent datum is effective. □

(9.2.4) Example. Let X be a connected smooth algebraic curve over a perfect field k. Let K denote the field of rational functions on X, $\overline{K}$ be a separable closure of K and let G denote the Galois group of $\overline{K}/K$. For each closed point x we choose a fixed extension $\bar{x}$ of the discrete valuation defined by x from K to $\overline{K}$. Let $I(\bar{x})$ denote the inertia group of $\bar{x}$ (cp. (8.1.9)). The quotient $\pi_1(X)$ of G by the closed normal subgroup generated by all $I(\bar{x})$ is the fundamental group of X. In other words, $\pi_1(X)$ is the Galois group of the maximal unramified subextension of $\overline{K}/K$.

Since we assumed X to be connected, all the stalks of a locally constant abelian sheaf F on $X_{\text{ét}}$ are isomorphic to a fixed abelian group A. Prove:

The locally constant abelian sheaves having stalks isomorphic to A correspond uniquely to the continuous homomorphisms

$$\pi_1(X) \to Aut(A).$$

(cp. [22], ch. XV).

(9.3) Constructible Sheaves

Let X be a topological space. A subset Z of X is called **retro-compact in** X, if for each quasi-compact open set $U \subset X$ the intersection $Z \cap U$ is again quasi-compact. Closed subsets of X are retro-compact in X.

A subset Z of X is called **constructible in** X if it is equal to a finite union of sets of the form $U \cap \complement V$ with $U, V \subset X$ open and retro-compact. The family of constructible subsets of X contains all retro-compact open subsets, it is stable under complements, finite intersections and finite unions, and it is minimal with respect to these properties (cp. [19], 0_{III}, 9.1.2., 9.1.3.). An open set $U \subset X$ is constructible if and only if it is retro-compact, a closed set $Y \subset X$ is constructible if and only if $\complement Y$ is retro-compact (cp. loc. cit., 9.1.5.).

If the topological space X is **noetherian**, every open subset of X is quasi-compact and hence retro-compact. Therefore the constructible subsets of X are precisely the finite unions of locally closed subsets.

If X is a scheme, a subscheme Z of X is called **constructible in** X if the underlying topological space of Z is constructible in the underlying topological space of X. If X is noetherian, each subscheme of X is constructible.

An abelian sheaf F on $X_{\text{ét}}$ is called **finite**, if all stalks $F_{\bar{x}}$ of F are finite. If Z is a subscheme of X and $i : Z \to X$ the canonical injection, then we denote the inverse image i^*F of an étale sheaf F on X also by F/Z.

(9.3.1) Definition. *An abelian sheaf F on $X_{\text{ét}}$ is called* **constructible** *if each affine open subset U of X has a decomposition into finitely many constructible reduced subschemes U_i of U such that F/U_i is locally constant and finite for all i.*

(9.3.2) Proposition. *i) Let F be an abelian sheaf on $X_{\text{ét}}$. Assume that X has a finite decomposition into constructible reduced subschemes X_i of X such that F/X_i is locally constant and finite for all i. Then F is constructible.*

Conversely, given a constructible sheaf F on $X_{\text{ét}}$, there exists a decomposition of X with the above properties, provided X is quasi-compact and quasi-separated.

ii) If $X = \bigcup U_i$ is an open covering of the scheme X, then an abelian sheaf F on $X_{\text{ét}}$ is constructible if and only if F/U_i is constructible for all i.

iii) If $f: Y \to X$ is a morphism of schemes and F is constructible on $X_{\text{ét}}$, then f^*F is constructible as well.

Proof: a) Assume X has a finite decomposition into constructible reduced subschemes X_i such that F/X_i is locally constant and finite. Let U be an affine open subset of X. The reduced subschemes $U_i = X_i \cap U$ of U are constructible ([19], 0_{III}, 9.1.8.) and yield a finite decomposition of U. Moreover, for each i the sheaf $F/U_i = (F/X_i)/U_i$ is locally constant and finite and therefore F is constructible.

b) Assume that X is quasi-compact and quasi-separated. Let F be an abelian sheaf on $X_{\text{ét}}$ so that for a fixed affine open covering $X = \bigcup U_i$ of X the restrictions F/U_i are all constructible. We claim that X has a decomposition of the kind mentioned in i). Obviously this will finish the proof of part i).

Since X is quasi-compact, we may assume the covering to be finite: $X = U_1 \cup \cdots \cup U_n$. By assumption each U_i has a decomposition $U_i = \bigcup U_i^j$ into finitely many constructible reduced subschemes U_i^j of U_i, such that

$F/U_i^j = (F/U_i)/U_i^j$ is locally constant and finite. The finitely many reduced subschemes

$$U_1^j, (X - U_1) \cap U_2^j, (X - (U_1 \cup U_2)) \cap U_3^j, \ldots$$

of X yield a decomposition of X. F induces on each of these subschemes a locally constant, finite sheaf. We are left to show that each subscheme above is constructible. Since X is quasi-separated, the intersection $U \cap V$ of two quasi-compact open subsets $U, V \subset X$ is again quasi-compact – note that $U \cap V = \Delta_X^{-1}(U \times_{\mathbb{Z}} V)$. Therefore all quasi-compact open subsets of X are constructible in X. In particular then $U_1, \ldots, U_n$ are constructible in X, and so are $X - U_1$, $X - (U_1 \cup U_2), \ldots$. By assumption all subschemes U_i^j are constructible in U_i. Since U_i is constructible in X, this implies ([19] 0_{III}, 9.1.8.) that all U_i^j are also constructible in X. Finally then all the subschemes $(X - U_1) \cap U_2^j, \ldots$ are constructible in X as intersections of constructible sets.

c) We prove ii): Let $X = \bigcup U_i$ be an open covering of the scheme X, and let F be an abelian sheaf on $X_{\text{ét}}$, such that F/U_i is constructible for all i. Let V be an affine open subset of X. Obviously then, there is an affine open covering $V = \bigcup V_j$ of V, such that all restrictions F/V_j are constructible. We apply part b) to this situation and obtain that F is constructible.

d) Let $f : Y \to X$ be a morphism of schemes and let F be a constructible sheaf on $X_{\text{ét}}$. To show that f^*F is constructible, it suffices by ii) to show the following: If U is an affine open subset of X, then $f^*F/f^{-1}(U)$ is constructible. Consider a finite decomposition $U = \bigcup U_i$ of U into constructible reduced subschemes U_i of U, such that F/U_i is locally constant and finite for all i. Then the inverse images $f^{-1}(U_i)$ give a finite decomposition of $f^{-1}(U)$ into constructible (cp. [20], 1.8.2.) reduced subschemes of $f^{-1}(U)$. Moreover, $f^*F/f^{-1}(U_i)$ is locally constant and finite for all i. Part i) implies that $f^*F/f^{-1}(U)$ is constructible. □

(9.3.3) Proposition. *Let G be a commutative étale group scheme on X. The sheaf G_X on $X_{\text{ét}}$ represented by G is constructible if and only if G is finitely presented on X.*

The proof of this proposition can be found in [2], exp. IX, 2.7.1.

Let R be a noetherian local ring, $X = spec(R)$, let x be the closed point of X and let $i : spec(k(x)) \to X$ be the canonical morphism. Let A be an abelian group and let A_x denote the constant sheaf on $spec(k(x))$ with value A. The sheaf i_*A_x on $X_{\text{ét}}$ is constructible (cp. (8.1.2)), but not locally constant.

Let k be a field, let $\bar{k}$ be a separable closure of k and let G denote the Galois group of $\bar{k}/k$. The category of constructible abelian sheaves on $spec(k)_{\text{ét}}$ is equivalent to the category of **finite** continuous G-modules.

(9.3.4) Example. Let X be a connected smooth algebraic curve over a perfect field k. Let K denote the field of rational functions on X and let $\overline{K}$ be a separable closure of K. For each closed point $x \in X$ let $\bar{x}$ denote an extension of the valuation defined by x from K to $\overline{K}$. The residue field $k(\bar{x})$ of $\bar{x}$ is an algebraic closure of $k(x)$, and we have a canonical isomorphism $Gal(k(\bar{x})/k(x)) \cong D(\bar{x})/I(\bar{x})$, where $D(\bar{x})$ and $I(\bar{x})$ denote the decomposition group and the inertia group of $\bar{x}$ respectively (cp. (8.1.9)). From the Decomposition Theorem (8.1.7) together with (8.1.9), (9.2.4) we obtain the following:

The constructible abelian sheaves on $X_{\text{ét}}$ are uniquely determined by the following data:

1⁰) A finite set S of closed points of X.
2⁰) For each point $x \in S$ a finite continuous $Gal(k(\bar{x})/k(x))$-module A_x.
3⁰) A finite continuous $\pi_1(X \setminus S)$-module A.
4⁰) For each point $x \in S$ a $Gal(k(\bar{x})/k(x))$-homomorphism $A_x \to A^{I(x)}$

§ 10. Étale Cohomology of Curves

(10.1) Skyscraper sheaves

(10.1.1) Lemma. *Let X be a noetherian sheaf on $X_{\text{ét}}$. The following are equivalent:*

i) For all non-closed points x of X we have $F_{\bar{x}} = 0$.

ii) We have

$$F \cong \bigoplus_x (i_x)_*(i_x)^*F,$$

where x runs through the closed points of X and $i_x : spec(k(x)) \to X$ denotes the canonical injection.

Proof: i) $\Longrightarrow$ ii): First we show the following: If X' is a finitely presented étale X-scheme and if $s \in F(X')$, then the support $supp(s) = \{x' \in X' \mid s_{\bar{x}'} \neq 0\}$ is a finite set of closed points of X'.

Certainly, $supp(s)$ is a closed subset of X'. Let $x' \in supp(s)$. Let $f : X' \to X$ denote the structure morphism of X'/X. By i) the point $x = f(x')$ is closed in X, hence $f^{-1}(x)$ is closed in X', and therefore x', being a closed point of $f^{-1}(x)$, is also closed in X'. Thus all points of $supp(s)$ are closed in X'. Since X is noetherian and since X' is a finitely presented X-scheme, the scheme X' is noetherian as well, hence $supp(s)$ is finite, since this is true in a noetherian scheme for any closed set consisting only of closed points.

The adjoint morphisms $F \to (i_x)_*(i_x)^*F$ induce a morphism

$$F \to \prod_x (i_x)_*(i_x)^*F$$

with x running through the closed points of X. For a closed point x of X the sheaf $(i_x)_*(i_x)^*F$ is concentrated in x and its stalk in $\bar{x}$ is isomorphic to $F_{\bar{x}}$. (cp. (6.4.2) and (5.2)). Let X' be a finitely presented étale X-scheme. What we proved at the beginning implies that the image of each $s \in F(X')$ under the map $F(X') \to ((i_x)_*(i_x)^*F)(X')$ is non-zero only for finitely many x. We conclude that the above morphism factors through the morphism

$$F \to \bigoplus_x (i_x)_*(i_x)^*F,$$

first on the restricted étale site, but then also on $X_{ét}$ (cp. (1.5.2)). That this in fact is an isomorphism can easily be seen by considering the stalks (cp. (6.6), i)), since the stalk functors commute with direct sums (cp. (5.2), i)).

ii) $\Longrightarrow$ i): The isomorphism

$$F \cong \bigoplus_x (i_x)_*(i_x)^*F$$

with x running through the closed points of X implies immediately that the stalk $F_{\bar{x}}$ is non-zero at most for closed points $x \in X$. □

An abelian sheaf F satisfying the equivalent conditions i) and ii) of (10.1.1) is called a **skyscraper sheaf** on $X_{ét}$.

(10.1.2) Lemma. *Let X be a noetherian scheme, such that the closed points $x \in X$ have separably closed residue fields $k(x)$. Then we have*

$$H^q(X, F) = 0 \quad \text{for} \quad q > 0$$

for all skyscraper sheaves F on $X_{ét}$.

Proof: By (10.1.1) we have an isomorphism

$$F \cong \bigoplus_x (i_x)_*(i_x)^* F,$$

where x runs through the closed points of X. Since X is noetherian, it is quasi-compact and quasi-separated, hence it follows from (1.5.3) that

$$H^q(X, F) \cong \bigoplus_x H^q(X, (i_x)_*(i_x)^* F).$$

For closed points x in X the morphism $i_x : spec(k(x)) \to X$ is a closed immersion. Therefore (6.4.2) ii) implies:

$$H^q(X, (i_x)_*(i_x)^* F) = H^q(spec(k(x)), (i_x)^* F).$$

By assumption $k(x)$ is separably closed, hence from (2.3) we finally get

$$H^q(spec(k(x)), (i_x)^* F) = 0$$

for $q > 0$. □

(10.1.3) Remark. If X is an algebraic scheme over a separably closed field k, then X satisfies the conditions of lemma (10.1.2).

We continue to assume that X is a noetherian scheme. Let K denote the ring of rational functions on X. As is well known (cp. [17], (8.1.8)) we have an isomorphism $K \cong \prod \mathcal{O}_{X,\xi}$, where ξ runs through the (finitely many) maximal points of X, hence through the generic points of the irreducible components of X. Let

$$j : spec(K) \to X$$

denote the canonical morphism, which is induced from the morphisms $spec(\mathcal{O}_{X,\xi}) \to X$. We are going to prove:

(10.1.4) Lemma. *If G is an abelian sheaf on $spec(K)_{ét}$, then we have*

$$(R^q j_*(G))_{\bar{\xi}} \cong \begin{cases} G_{\bar{\xi}} & \text{for } q = 0 \\ 0 & \text{for } q > 0 \end{cases}$$

for each maximal point ξ of X.

Proof: The morphism $j : spec(K) \to X$ is quasi-compact and quasi-separated. Therefore by (6.4.1):

$$(R^q j_*(G))_{\bar{\xi}} \cong H^q(spec(K) \times_X X(\bar{\xi}), \overline{G}).$$

Here $X(\bar{\xi})$ denotes the strict localization of X in the geometric point $\bar{\xi}$ and $\overline{G}$ denotes the inverse image of G under $spec(K) \times_X X(\bar{\xi}) \to spec(K)$. The scheme $spec(K) \times_X X(\bar{\xi})$ can be canonically identified with $X(\bar{\xi})$. Now, the scheme $X(\bar{\xi})$ is the affine scheme of the strict henselization $\mathcal{O}^{hs}_{X,\xi}$ of $\mathcal{O}_{X,\xi}$ with respect to $k(\xi) \to k(\bar{\xi})$. Since $\mathcal{O}_{X,\xi}$ is artinian, the same is true for $\mathcal{O}^{hs}_{X,\xi}$ (cp. (6.1.9)). Therefore, by (8.1.3), the morphism $spec(k(\bar{\xi})) \to X(\bar{\xi})$ induces an equivalence between the categories of abelian sheaves on $X(\bar{\xi})_{ét}$ and on $spec(k(\bar{\xi}))_{ét}$. From this and (2.3) the claim follows. □

The next two propositions will show how the statements (10.1.2) and (10.1.4) can be used to compute the cohomology groups of arbitrary abelian sheaves F on $X_{ét}$, in case X is a **1-dimensional** noetherian scheme.

(10.1.5) Proposition. *Let X be a 1-dimensional noetherian scheme such that the residue fields $k(x)$ of closed points $x \in X$ are separably closed. Let K denote the ring of rational functions on X and let $j : spec(K) \to X$ denote the canonical morphism. For any abelian sheaf F on $X_{ét}$ the adjoint morhism $\rho : F \to j_*j^*F$ induces isomorphisms*

$$H^q(X, F) \cong H^q(X, j_*j^*F)$$

in dimensions $q \geq 2$.

Proof: If ξ is a maximal point of X, the results in (10.1.4) and (5.2) iii) imply that $(j_*j^*F)_{\bar{\xi}} \cong (j^*F)_{\bar{\xi}} \cong F_{\bar{\xi}}$, hence that the kernel and the cokernel of the adjoint morphism $\rho : F \to j_*j^*F$ are skyscrapersheaves on $X_{ét}$. If we look at the long cohomology sequences attached to the exact sequences

$$0 \to ker(\rho) \to F \to im(\rho) \to 0$$
$$0 \to im(\rho) \to j_*j^*F \to coker(\rho) \to 0,$$

we obtain the isomorphism $H^q(X, F) \cong H^q(X, j_*j^*F)$ for $q \geq 2$, since by (10.1.2) the cohomology groups $H^q(X, ker(\rho))$ and $H^q(X, coker(\rho))$ vanish for $q \geq 1$. □

Proposition (10.1.5) reduces the computation of the cohomology groups $H^q(X, F)$ in dimensions $q \geq 2$ to the computation of the cohomology with values in sheaves of the form j_*G, where G is an abelian sheaf on $spec(K)_{ét}$. We have the following general information about these cohomology groups $H^q(X, j_*G)$:

(10.1.6) Proposition. *Let X be a 1-dimensional noetherian scheme, such that the residue fields $k(\bar{x})$ of the closed points $x \in X$ are separably closed. Let K denote the ring of rational functions on X and let $j : spec(K) \to X$ denote the canonical morphism. For all abelian sheaves G on $spec(K)_{ét}$ we have the following exact sequences:*

$$0 \to H^1(X, j_*G) \to H^1(spec(K), G) \to \bigoplus_x H^1(spec(K) \times_X X(x), G)$$
$$\to H^2(X, j_*G) \to H^2(spec(K), G) \to \bigoplus_x H^2(spec(K) \times_X X(x), G)$$
$$\to H^3(X, j_*G) \to \cdots$$

Here x runs through the closed points of X and $X(x)$ denotes the strict localization of X in x.

Proof: Consider the Leray spectral sequence (1.4.3):

$$E_2^{pq} = H^p(X, R^q j_*(G)) \Longrightarrow E^{p+q} = H^{p+q}(spec(K), G).$$

By (10.1.4) $R^q(j_*G)$ is a skyscrapersheaf on $X_{ét}$ for $q > 0$, hence by (10.1.2):

$$E_2^{pq} = H^p(X, R^q(j_*G)) = 0 \quad \text{for } p > 0,\ q > 0.$$

Therefore the spectral sequence $E_2^{pq} \Longrightarrow E^{p+q}$ induces an exact sequence (cp. [5], XV, 5.):

$$\begin{aligned} 0 \to E_2^{1,0} &\to E^1 \to E_2^{0,1} \\ &\to E_2^{2,0} \to E^2 \to E_2^{0,2} \\ &\to E_2^{3,0} \to \cdots \end{aligned}$$

Using (10.1.1) and (6.4.1) we obtain for the terms $E_2^{0,q}$ for $q > 0$:

$$\begin{aligned} E_2^{0,q} &= H^0(X, R^q j_*(G)) \\ &\cong H^0(X, \bigoplus_x (i_x)_*(i_x)^* R^q j_*(G)) \\ &\cong \bigoplus_x H^0(spec(k(x)), (i_x)^* R^q j_*(G)) \\ &\cong \bigoplus_x (R^q j_*(G))_x \\ &\cong \bigoplus_x H^q(spec(K) \times_X X(x), G). \end{aligned}$$

Here, as in the statement of the proposition, we denoted the inverse image of G under $spec(K) \times_X X(x) \to spec(K)$ again by G. □

(10.2) The Cohomological Dimension of Algebraic Curves

Let X be a noetherian scheme of dimension 1 and let K be the ring of rational functions on X. (8.1.3) implies the following result on the cohomological ℓ-dimension $cd_\ell(K)$ (cp. (9.1.7)) of K, ℓ a prime number:

$$cd_\ell(K) = cd_\ell(K_{red}) = \max_\xi cd_\ell(k(\xi)),$$

where ξ runs through the maximal points of X.

(10.2.1) Proposition. *Let X be a 1-dimensional noetherian scheme, such that the residue fields of the closed points are separably closed. Let K denote the ring of rational functions on X and let $j : spec(K) \to X$ denote the canonical morphism. If $cd_\ell(K) \le r$, we have the following result for all ℓ-torsion sheaves G on $spec(K)_{\text{ét}}$:*

i) $H^q(X, j_*G) = 0$ for $q > r+1$.

ii) There is an exact sequence

$$\begin{aligned} 0 \to H^1(X, j_*G) \to H^1(spec(K), G) &\to \bigoplus_x H^1(spec(K) \times_X X(x), G) \\ &\to H^2(X, j_*G) \to \cdots \to H^{r+1}(X, j_*G) \to 0. \end{aligned}$$

Proof: We use the exact sequence of proposition (10.1.6). Since inverse images of ℓ-torsion sheaves are again ℓ-torsion sheaves (cp. (9.1.6)), the statements i) and ii) are consequences of the following

(10.2.2) Lemma. *If X is a noetherian scheme and K is the ring of rational functions on X, we have*

$$cd_\ell(spec(K) \times_X X(\bar{x})) \leq cd_\ell(K)$$

for all geometric points $\bar{x}$ of X.

Proof of (10.2.2): The fibre of $spec(K) \times_X X(\bar{x}) \to spec(K)$ over a point ξ of K is equal to a finite sum of spectra of separable algebraic field extensions of $k(\xi)$ (cp. [20], 18.8.12.). Now Galois cohomology implies that for an algebraic field extension k'/k we have $cd_\ell(k') \leq cd_\ell(k)$ (cp. [32], ch. II, 4.1., prop. 10). Therefore the cohomological ℓ-dimension of the fibre of $spec(K) \times_X X(\bar{x}) \to spec(K)$ over the point ξ of $spec(K)$ is less or equal to $cd_\ell(k(\xi))$. Hence (cp. (8.1.3))

$$cd_\ell(spec(\mathcal{O}_{X,\xi}) \times_X X(\bar{x})) \leq cd_\ell(\mathcal{O}_{X,\xi}). \qquad \square$$

(10.2.3) Theorem. *Let X be a 1-dimensional noetherian scheme, such that the residue fields of closed points are separably closed. Let K denote the ring of rational functions on X. Then*

$$cd_\ell(X) \leq cd_\ell(K) + 1$$

and also (cp. (9.1.7))

$$cd(X) \leq cd(K) + 1.$$

Proof: If $cd_\ell(K) = \infty$, there is nothing to prove. Assume that $cd_\ell(K)$ is finite and take $q > cd_\ell(K) + 1$. Let F be a ℓ-torsion sheaf on $X_{\text{ét}}$. We have to show that $H^q(X, F) = 0$. Let $j : spec(K) \to X$ denote the canonical morphism. Since $q \geq 2$, (10.1.5) implies that

$$H^q(X, F) \cong H^q(X, j_*j^*F).$$

The sheaf j^*F is a ℓ-torsion sheaf on $spec(K)_{\text{ét}}$ (cp. (9.1.6)) and therefore by (10.2.1), i):

$$H^q(X, j_*j^*F) = 0 \quad \text{for} \quad q > cd_\ell(K) + 1. \qquad \square$$

(10.2.4) Theorem. *Let X be an algebraic curve over a field k (cp. [12], 7.4.2.). Then*

$$cd_\ell(X) \leq 2 + cd_\ell(k).$$

In particular, if k is separably closed, we have

$$cd_\ell(X) \leq 2$$

and hence

$$cd(X) \leq 2.$$

Proof: Let us first assume that k is separably closed. Then X satisfies the assumptions of the previous theorem and hence

$$cd_\ell(X) \leq cd_\ell(K) + 1,$$

where K is the ring of rational functions on X. As we mentioned at the beginning of this section, we have

$$cd_\ell(K) = \max_\xi cd_\ell(k(\xi)),$$

where ξ runs through the maximal points of X. The field $k(\xi)$ has transcendence degree 1 over the separably closed field k. Therefore by Tsen's Theorem (cp. [32], ch. II, § 3) we have $cd_\ell(k(\xi)) \leq 1$. Hence $cd_\ell(K) \leq 1$ and therefore $cd_\ell(X) \leq 2$.

Now let k be an arbitrary field. We may assume that $cd_\ell(k)$ is finite and we have to show that $H^n(X, F) = 0$ for all ℓ-torsion sheaves F on $X_{\text{ét}}$ and $n > 2 + cd_\ell(k)$. We use the Artin spectral sequence (7.2): Let $\bar{k}$ be a separable closure of k, $G = Gal(\bar{k}/k)$, $\overline{X} = X \times_k \bar{k}$ and let $\overline{F}$ be the inverse image of F under $\overline{X} \to X$. The spectral sequence reads:

$$E_2^{p,q} = H^p(G, H^q(\overline{X}, \overline{F})) \Longrightarrow E^{p+q} = H^{p+q}(X, F).$$

By (9.1.3) (cp. (9.1.6)) the sheaf $\overline{F}$ is a ℓ-torsion sheaf on $\overline{X}_{\text{ét}}$ and by (9.1.2) (cp. (9.1.6)) the cohomology groups $H^q(\overline{X}, \overline{F})$ are ℓ-torsion groups for all q. We just proved that $cd_\ell(\overline{X}) \leq 2$. Therefore for $n > 2 + cd_\ell(k)$ we have:

$$E_2^{p,q} = H^p(G, H^q(\overline{X}, \overline{F})) = 0, \text{ if } p + q = n.$$

But this implies $E^n = H^n(X, F) = 0$ for $n > 2 + cd_\ell(k)$. (Note that for an arbitrary spectral sequence $E_2^{p,q} \Longrightarrow E^{p+q}$, the vanishing of E_2^{pq} for $p + q = n$ implies first of all that $E_\infty^{pq} = 0$ for $p + q = n$ and then $E^n = 0$, as is easily seen from the definition of spectral sequences in 0, (2.3)). □

(10.2.5) Remark. If X is an **affine** algebraic curve over a separably closed field k, the stronger result

$$cd_\ell(X) \leq 1$$

holds (as well as $cd_\ell(X) \leq 1 + cd_\ell(k)$ for general fields k). A proof can be found in [2], exp. IX, 5.7. and exp. X, 5.2.

(10.3) The Groups $H^q(X, (\mathbb{G}_m)_X)$ and $H^q(X, (\mu_n)_X)$

For an arbitrary scheme X we have $H^0(X, (\mathbb{G}_m)_X) = \Gamma(X, \mathcal{O}_X)^*$ and $H^1(X, (\mathbb{G}_m)_X) \cong Pic(X)$ (cp. (4.3.1)). Moreover we proved (cp. (9.1.5)) that the groups $H^q(X, (\mathbb{G}_m)_X)$ are torsion for $q \geq 2$, provided the scheme X is **regular** and noetherian. We now show:

(10.3.1) Proposition. *Let X be a 1-dimensional noetherian scheme, such that the residue fields of closed points are separably closed. Let K be the ring of rational functions on X. Then we have: For $q \geq 2$ the groups $H^q(X, (\mathbb{G}_m)_X)$ are torsion with*

$$H^q(X, (\mathbb{G}_m)_X)(\ell) = 0$$

for all prime numbers ℓ, which are invertible on X and satisfy $cd_\ell(K) \leq 1$.

Proof: Let $j : spec(K) \to X$ denote the canonical morphism. By (10.1.5) we have

$$H^q(X, (\mathbb{G}_m)_X) \cong H^q(X, j_*j^*(\mathbb{G}_m)_X)$$

for $q \geq 2$. If we look at the stalks in the geometric points $\bar{\xi}$ of $spec(K)$, then we realize that the sheaf $j^*(\mathbb{G}_m)_X$ gets identified with the sheaf $(\mathbb{G}_m)_K$ via the canonical morphism $j^*(\mathbb{G}_m)_X \to (\mathbb{G}_m)_K$. Here we used $(\mathbb{G}_m)_K$ to abbreviate $(\mathbb{G}_m)_{spec(K)}$. Therefore we have to show that for $q \geq 2$ the groups $H^q(X, j_*(\mathbb{G}_m)_K)$ are torsion and that their ℓ-torsion parts vanish for the prime numbers ℓ mentioned in the proposition.

We consider the exact sequence (10.1.6) for the sheaf $(\mathbb{G}_m)_K$ on $spec(K)_{\text{ét}}$:

$$\begin{aligned} 0 &\to H^1(X, j_*(\mathbb{G}_m)_K) \to H^1(spec(K), (\mathbb{G}_m)_K) \\ &\to \bigoplus_x H^1(spec(K) \times_X X(x), \pi^*(\mathbb{G}_m)_K) \to H^2(X, j_*(\mathbb{G}_m)_K) \\ &\to \cdots \end{aligned}$$

Here $X(x)$ is the strict localization of X in x and $\pi : spec(K)\times_X X(x) \to spec(K)$ is the projection onto the first factor. The sequence shows that it suffices to prove that the groups $H^q(spec(K), (\mathbb{G}_m)_K)$ and $H^q(spec(K)\times_X X(x), \pi^*(\mathbb{G}_m)_K)$ are torsion for $q \geq 1$ and their ℓ-torsion parts vanish for the predicted values of ℓ.

That the groups $H^q(spec(K), (\mathbb{G}_m)_K)$ are torsion for $q \geq 1$ follows from (8.1.3), (2.2) and Galois cohomology ([32], ch. I, 2.2., cor. 3). Furthermore, $H^1(spec(K), (\mathbb{G}_m)_K) \cong Pic(K) = 0$. Assume now that ℓ is a prime number which is invertible on X, and that $cd_\ell(K) \leq 1$. Then ℓ is also invertible on $spec(K)$, and we have an exact sequence (cp. (4.4.1))

$$0 \to (\mu_\ell)_K \to (\mathbb{G}_m)_K \to (\mathbb{G}_m)_K \to 0.$$

Since $H^q(spec(K), (\mu_\ell)_K) = 0$ for $q \geq 2$, the long exact cohomology sequence implies that the maps

$$H^q(spec(K), (\mathbb{G}_m)_K) \xrightarrow{\ell} H^q(spec(K), (\mathbb{G}_m)_K)$$

are isomorphisms for $q \geq 2$. But this means that $H^q(spec(K), (\mathbb{G}_m)_K)(\ell) = 0$ for $q \geq 2$.

The sheaf $\pi^*(\mathbb{G}_m)_K$ can be identified with the sheaf $(\mathbb{G}_m)_{spec(K)\times_X X(x)}$ and we conclude as above that the groups $H^q(spec(K)\times_X X(x), \pi^*(\mathbb{G}_m)_K)$ are torsion for $q \geq 2$, that $H^1(spec(K)\times_X X(x), \pi^*(\mathbb{G}_m)_K) = 0$ and that $H^q(spec(K)\times_X X(x), \pi^*(\mathbb{G}_m)_K)(\ell) = 0$ for $q \geq 2$. □

(10.3.2) Corollary. *If X is an algebraic curve over a separably closed field k of characteristic p, then*

$$H^q(X, (\mathbb{G}_m)_X)(\ell) = 0$$

for $q \geq 2$ and all prime numbers $\ell \neq p$. Hence for $q \geq 2$ the groups $H^q(X, (\mathbb{G}_m)_X)$ are p-torsion groups.

Proof: The conditions of (10.3.1) are satisfied for all primes $\ell \neq \mathrm{char}\, k = p$. □

(10.3.3) Proposition. *Let X be a 1-dimensional noetherian scheme, such that the residue fields of closed points are separably closed. Let n be a natural number invertible on X, such that $cd_\ell(K) \leq 1$ holds for all primes ℓ dividing n. Then*

$$H^q(X, (\mu_n)_X) = 0 \quad \text{for} \quad q > 2,$$

and for $q \leq 2$ the cohomology of $(\mu_n)_X$ is given by the following exact sequence:

$$\begin{aligned} 0 &\to H^0(X, (\mu_n)_X) \to \Gamma(X, \mathcal{O}_X^*) \xrightarrow{n} \Gamma(X, \mathcal{O}_X^*) \\ &\to H^1(X, (\mu_n)_X) \to Pic(X) \xrightarrow{n} Pic(X) \\ &\to H^2(X, (\mu_n)_X) \to 0. \end{aligned}$$

Proof: For all primes ℓ dividing n we have $cd_\ell(X) \leq cd_\ell(K) + 1 \leq 2$ by theorem (10.2.3) and the assumption on $cd_\ell(K)$. Hence $H^q(X, (\mu_n)_X(\ell)) = 0$ for $q > 2$ (cp. (9.1.7)). If a prime ℓ' is relatively prime to n, we have $(\mu_n)_X(\ell') = 0$, and hence $H^q(X, (\mu_n)_X)(\ell') = H^q(X, (\mu_n)_X(\ell')) = 0$ for all $q \geq 0$ (cp. loc. cit.). Therefore $H^q(X, (\mu_n)_X) = 0$ for $q > 2$.

Since n is invertible on X, we have the exact Kummer sequence (4.4.1)

$$0 \to (\mu_n)_X \to (\mathbb{G}_m)_X \xrightarrow{n} (\mathbb{G}_m)_X \to 0.$$

The corresponding cohomology sequence induces the exact sequence stated in the proposition, since by (10.3.1) we have $H^2(X, (\mathbb{G}_m)_X)(\ell) = 0$ for all ℓ / n, and therefore the map $H^2(X, (\mathbb{G}_m)_X) \xrightarrow{n} H^2(X, (\mathbb{G}_m)_X)$ is injective. □

(10.3.4) Theorem. *If X is an algebraic curve over a separably closed field k, and if n is relatively prime to the characteristic of k, we have*

$$H^q(X, (\mu_n)_X) = 0 \quad \text{for} \quad q > 2$$

and an exact sequence

$$\begin{aligned} 0 &\to H^0(X, (\mu_n)_X) \to \Gamma(X, \mathcal{O}_X^*) \xrightarrow{n} \Gamma(X, \mathcal{O}_X^*) \\ &\to H^1(X, (\mu_n)_X) \to Pic(X) \xrightarrow{n} Pic(X) \\ &\to H^2(X, (\mu_n)_X) \to 0. \end{aligned}$$

(10.3.5) Theorem. *Let X be a connected complete algebraic curve over a separably closed field k. Then we have for all n, relatively prime to the characteristic of k:*

$$\begin{aligned} H^0(X,(\mu_n)_X) &\cong \mu_n(k) \\ H^1(X,(\mu_n)_X) &\cong {}_nPic(X) \\ H^2(X,(\mu_n)_X) &\cong Pic(X)_n \\ H^q(X,(\mu_n)_X) &= 0 \quad \text{for } q > 2. \end{aligned}$$

(10.3.6) Remark. Under the assumptions in (10.3.4) and (10.3.5) we have an isomorphism $(\mu_n)_X \cong (\mathbb{Z}/n\mathbb{Z})_X$, so that both theorems can also be viewed as results on the cohomology of the constant sheaf $(\mathbb{Z}/n\mathbb{Z})_X$.

If in (10.3.5) we assume in addition that X is **smooth**, we have the well known fact that ${}_nPic(X) \cong (\mathbb{Z}/n\mathbb{Z})^{2g}$, where g is the genus of X. Furthermore: $\mu_n(k) \cong \mathbb{Z}/n\mathbb{Z}$ and $Pic(X)_n \cong \mathbb{Z}/n\mathbb{Z}$, so that the cohomology groups of $(\mu_n)_X \cong (\mathbb{Z}/n\mathbb{Z})_X$ have ranks $1, 2g, 1$ as $(\mathbb{Z}/n\mathbb{Z})$-modules.

(10.4) The Finiteness Theorem for Constructible Sheaves

Let X be a scheme and let p be a prime number. We call a finite (cp. (9.3)) abelian sheaf F on $X_{\text{ét}}$ **of order prime to** p, if each stalk $F_{\bar{x}}$ of F has order prime to p.

(10.4.1) Theorem *(Finiteness Theorem).*

Let X be a smooth algebraic curve over a separably closed field k of characteristic p. If F is a constructible abelian sheaf on $X_{\text{ét}}$ of order prime to p, then the cohomology groups $H^q(X,F)$ are finite for all $q \geq 0$.

Proof: Let $\bar{k}$ be an algebraic closure of k, let $\overline{X} = X \times_k \bar{k}$ and let $f : \overline{X} \to X$ denote the canonical projection. By descent theory (cp. [16], exp. IX, 4.11.) the morphism $f_{\text{ét}} : X_{\text{ét}} \to \overline{X}_{\text{ét}}$ is an equivalence of topologies. Therefore, for any abelian sheaf F on $X_{\text{ét}}$, we have an isomorphism $H^q(X,F) \cong H^q(\overline{X},\overline{F})$, where $\overline{F}$ denotes the inverse image of F under $\overline{X} \to X$. If F is constructible and of order prime to p, the same holds for $\overline{F}$ (cp. (9.3.2) and (5.2)), and hence we may assume that k is algebraically closed.

Furthermore, we may assume that the curve X is connected.

We are going to reduce the Finiteness Theorem in several steps to the finiteness of the groups $H^q(X,(\mathbb{Z}/n\mathbb{Z})_X)$ for $(n,p)=1$ and for complete smooth connected curves X, which we proved in (10.3.6).

a) Let X be a smooth algebraic curve over the algebraically closed field k. First we want to show that it suffices to prove the Finiteness Theorem for **locally constant** sheaves.

Let F be a constructible abelian sheaf on $X_{\acute{e}t}$ of order prime to p. By (9.3.2), i) there is a non-empty open subset U of X, such that F/U is locally constant. Let $Y = X \smallsetminus U$, equipped with the reduced subscheme structure of X. We claim that the relative cohomology groups $H^q_Y(X,F)$ are finite for $q \geq 0$. The exact relative cohomology sequence (cp. (8.3.2))

$$\cdots \to H^q_Y(X,F) \to H^q(X,F) \to H^q(U,F) \to H^{q+1}_Y(X,F) \to \cdots$$

would then imply that $H^q(X,F)$ is finite as soon as $H^q(U,F)$ is finite. To prove the finiteness of the relative cohomology groups $H^q_Y(X,F)$ we consider the spectral sequence (8.3.1)

$$E_2^{pq} = H^p(Y, R^q i^!(F)) \Longrightarrow E^{p+q} = H^{p+q}_Y(X,F).$$

It suffices to show the finiteness of the initial terms $E_2^{pq} = H^p(Y,R^q i^!(F))$. This follows from the definition of spectral sequences and from 0, (2.3.1). The scheme Y is equal to a finite sum of copies of the scheme $spec(k)$. Therefore, to prove the finiteness of $H^p(Y,R^q i^!(F))$, it suffices to show that the sheaves $R^q i^!(F)$ are finite on $Y_{\acute{e}t}$. We use the results from (8.2.4), namely the exact sequence

$$0 \to i_* i^!(F) \to F \to j_* j^* F \to i_* R^1 i^!(F) \to 0$$

and the formula

$$i_* R^q i^!(F) \cong R^{q-1} j_*(j^* F) \quad \text{for } q \geq 2.$$

Here $j: U \to X$ denotes the canonical injection. Since by (6.4.2) $(i_*G)_{\bar{y}} \cong G_{\bar{y}}$ holds for abelian sheaves on $Y_{\acute{e}t}$ and points $y \in Y$, we conclude from the exact sequence that $i^!(F)$ is finite. We also conclude that $R^q i^!(F)$ will be finite for $q > 0$ if we can show that $R^q j_*(j^* F)$ is finite for $q \geq 0$.

This will follow from Lemma (10.4.2), applied to the locally constant sheaf $j^*F = F/U$ on $U_{\text{ét}}$ with order prime to p, and will finish the first reduction step.

b) Next we show that it suffices to prove the Finiteness Theorem for **constant** sheaves.

Let F be a locally constant finite abelian sheaf on $X_{\text{ét}}$ of order prime to p. Let A denote the typical stalk of F. Then F determines a continuous action of the fundamental group $\pi_1(X)$ on A, which factors through a finite quotient G of $\pi_1(X)$ (cp. (9.2.4)). There exists a connected smooth algebraic curve X' over k, which is Galois with group G, finite and étale over X (cp. [18], 7.4.). By construction $\pi_1(X')$ acts trivially on A, hence $F/X' \cong A_{X'}$ is constant. Now consider the Artin spectral sequence (7.1)

$$E_2^{pq} = H^p(G, H^q(X', F/X')) \Longrightarrow E^{p+q} = H^{p+q}(X, F).$$

If we assume the Finiteness Theorem for constant finite sheaves of order prime to p, then the initial terms $E_2^{pq} = H^p(G, H^q(X', F/X'))$ of the spectral sequence are finite, and therefore the limit terms $E^n = H^n(X, F)$ are finite as well.

c) Let us show now that for the proof of the theorem for constant sheaves we may assume that the curve X is **complete**.

Let U be a non-complete (and hence affine (cp. [18], 7.4.12.)) connected smooth curve over k and let $j : U \to X$ be the injection into the smooth completion X of U (cp. [18], 7.4.11.). If A_U is a constant sheaf on $U_{\text{ét}}$, then we have $j_*(A_U) \cong A_X$ (cp. the second remark following (8.1.2)). Hence we are left to show: If A is finite of order prime to p, then the finiteness of all groups $H^q(X, j_*(A_U))$ implies the finiteness of the groups $H^q(U, A_U)$. Consider the Leray spectral sequence

$$E_2^{pq} = H^p(X, R^q j_*(A_U)) \Longrightarrow E^{p+q} = H^{p+q}(U, A_U).$$

Again we show that the initial terms E_2^{pq} are finite. By assumption the terms $E_2^{p,0} = H^p(X, j_*(A_U))$ are finite. Lemma (10.4.2), which we used already in part a), will imply that the sheaves $R^q j_*(A_U)$ on $X_{\text{ét}}$ are finite for $q \geq 0$. Moreover, for $q > 0$ these sheaves $R^q j_*(A_U)$ are concentrated on the finite set $Y = X \setminus U$. This can either be seen by looking at the stalks using (6.4.1) and (6.2.3) or through the relation (cp. (8.2.4))

$$R^q j_*(A_U) \cong R^q j_*(j^* A_X) \cong i_* R^{q+1} i^!(A_X),$$

where $i : Y \to X$ is the injection of the reduced subscheme Y into X. In any case it follows that the terms $E_2^{pq} = H^p(X, R^q j_*(A_U))$ for $q > 0$ are finite as well. This completes the proof of reduction step c) up to Lemma (10.4.2).

d) Assume now that X is complete and that A is a finite abelian group of order prime to p. We have to show that $H^q(X, A_X)$ is finite for $q \geq 0$. But this follows from (10.3.6), since the sheaf A_X is isomorphic to a finite direct sum of sheaves of the form $(\mathbb{Z}/\ell^r\mathbb{Z})_X$ with primes $\ell \neq p$. □

We still have to prove the following

(10.4.2) Lemma. *Let $j : U \to X$ be an open immersion of connected algebraic curves over an algebraically closed field k. Let F be a locally constant finite abelian sheaf on $U_{\text{ét}}$ of order prime to $p = \mathrm{char}(k)$. Then the sheaves $R^q j_*(F)$ are finite.*

Proof: By (6.4.1) we have the following isomorphism for the stalks $(R^q j_*(F))_{\bar{x}}$ in points $x \in X$:

$$(R^q j_*(F))_{\bar{x}} \cong H^q(U \times_X X(\bar{x}), F),$$

where $X(\bar{x})$ denotes the strict localization of X in $\bar{x}$.

Let $x \in U$. Then $U \times_X X(\bar{x}) = X(\bar{x})$ and from (6.2.3) we obtain:

$$(R^q j_*(F))_{\bar{x}} \cong \begin{cases} F_{\bar{x}} & \text{for } q = 0 \\ 0 & \text{for } q > 0. \end{cases}$$

Assume now $x \notin U$. Then $U \times_X X(\bar{x})$ is equal to the spectrum of the quotient field of the henselization $\mathcal{O}_{X,x}^h$ of $\mathcal{O}_{X,x}$, and the finiteness of the stalk $(R^q j_*(F))_{\bar{x}}$ results from

(10.4.3) Lemma. *Let R be a henselian discrete valuation ring with algebraically closed residue field k and quotient field K. Let $p = char\, k = char\, K$. Let $\overline{K}$ be a separable closure of K and let $G = Gal(\overline{K}/K)$. Then the cohomology groups $H^q(G, A)$ with coefficients in finite continuous G-modules A of order prime to p are finite (and $= 0$ for $q \geq 2$).*

Proof of (10.4.3): We may assume that R is complete (cp. [2], exp. X, 2.2.1.). Let n be prime to p. The cohomology groups with values in μ_n are as follows:

$$\begin{aligned} &H^0(G, \mu_n) = \mu_n(K) \\ &H^1(G, \mu_n) \cong K^*/K^{*n} = \mathbb{Z}/n\mathbb{Z} \\ &H^q(G, \mu_n) = 0 \quad \text{for } q \geq 2. \end{aligned}$$

Note that $cd(K) = 1$ (cp. [32], II, prop. 12). In particular, the groups $H^q(G, \mu_n)$ are finite, and this remains true if we replace G by any open normal subgroup of G. Now given a finite continuous G-module A of order prime to p, we choose an open normal subgroup H of G such that A is isomorphic as an H-module to a finite direct sum of modules of the form μ_n, n prime to p. Then $H^q(H, A)$ is finite, and the Hochschild-Serre spectral sequence (cp. I, (3.7.9,10))

$$H^p(G/H, H^q(H, A)) \Longrightarrow H^{p+q}(G, A)$$

implies that the groups $H^n(G, A)$ are finite as well. □

Remark. If we do not assume in Theorem (10.4.1) that the constructible sheaf F has order prime to p, the groups $H^q(X, F)$ are not necessarily finite. E.g. if $X = spec(k[t])$ is the affine line, then the Artin-Schreier theory (cp. (4.2.3)) implies that

$$H^1(X, (\mathbb{Z}/p\mathbb{Z})_X) \cong k[t]/\wp k[t],$$

and $k[t]/\wp k[t]$ is not finite. However, if we assume that X is complete, the Finiteness Theorem remains true for arbitrary constructible sheaves (cp. (11.4.2)).

§ 11. General Theorems in Étale Cohomology Theory

In the following we mention without proofs some of the most important theorems in étale cohomology.

(11.1) The Comparison Theorem with Classical Cohomology

Let X be an arbitrary topological space. Consider the category of local homeomorphisms $f : U \to X$. These are continuous maps $f : U \to X$ such that each $x \in U$ has an open neighbourhood U_x in U, which is mapped homeomorphic under f to a neighbourhood of $f(x)$. We define a topology on this category by taking as coverings families $\{U_i \xrightarrow{f_i} U\}$ with $\bigcup f_i(U_i) = U$. This topology is denoted by X_{cl}. Since open embeddings are locally homeomorphic, we have a natural morphism

$$i : X \to X_{cl}$$

of topologies. This morphism satisfies the conditions i) – iii) of the Comparison Lemma I, (3.9.1). Hence the functors

$$i^s : \widetilde{X}_{cl} \to \widetilde{X}$$
$$i_s : \widetilde{X} \to \widetilde{X}_{cl}$$

are quasi-inverse equivalences between the categories $\widetilde{X}_{cl}$ and $\widetilde{X}$ of abelian sheaves on X_{cl} and X respectively. Therefore, for the study of the cohomology theory of abelian sheaves, we can replace X by the site X_{cl}.

Assume now that X is a locally algebraic scheme over the field $\mathbb{C}$ of complex numbers. The set $X(\mathbb{C})$ of $\mathbb{C}$-rational points of X can be equipped in a well known way with the structure of a topological space: a basis of the topology is given by the sets

$$\{x \in U(\mathbb{C}) \mid |f_i(x)| < \varepsilon \quad \text{for } i = 1, \ldots, r\},$$

where $U \subset X$ is Zariski-open, $f_1, \ldots, f_r \in \Gamma(U, \mathcal{O}_X)$ and $\varepsilon > 0$ arbitrary.

Now the following holds: If $f : X' \to X$ is an étale morphism of schemes, then the induced map $X'(\mathbb{C}) \to X(\mathbb{C})$ is a local homeomorphism. Furthermore, if $X' \to X$ is surjective, the same holds for $X'(\mathbb{C}) \to X(\mathbb{C})$. Hence the functor $X' \to X'(\mathbb{C})$ is a morphism of topologies

$$\varepsilon : X_{\text{ét}} \to X_{cl},$$

where X_{cl} stands for $X_{cl}(\mathbb{C})$.

(11.1.1) Comparison Theorem. *Let X be a smooth locally algebraic $\mathbb{C}$-scheme. Then:*

i) The functors ε^s and ε_s induce mutually quasi-inverse equivalences between the category of locally constant finite abelian sheaves on $X_{\text{ét}}$ and the category of locally constant finite abelian sheaves on X_{cl}.

ii) If F is a locally constant finite abelian sheaf on X_{cl}, then we have $R^q\varepsilon^s(F) = 0$ for $q > 0$, and therefore canonical isomorphisms

$$H^p(X_{\text{ét}}, \varepsilon^s(F)) \cong H^p(X_{cl}, F).$$

In particular,

$$H^p(X_{\text{ét}}, \mathbb{Z}/n\mathbb{Z}) \cong H^p(X_{cl}, \mathbb{Z}/n\mathbb{Z}).$$

The proof of this theorem can be found in [2], exp. XI. For a more general version, without assuming $X/\mathbb{C}$ to be smooth, see [2], exp. XVI, 4.

(11.2) The Cohomological Dimension of Algebraic Schemes

(11.2.1) Theorem. *Let X be an algebraic scheme over a field k of characteristic p. Then*

$$\begin{cases} cd_\ell(X) \leq 2 \dim X + cd_\ell(k) & \text{for } \ell \neq p \\ cd_p(X) \leq \dim X + 1. \end{cases}$$

(11.2.2) Corollary. *If X is an algebraic scheme over a separably closed field k, then*

$$cd(X) \leq 2 \dim X.$$

For a proof of (11.2.1) see [2], exp. X, 4.3. and 5.2.

(11.2.3) Theorem. *Let X be an* **affine** *algebraic scheme over a separably closed field k. Then*

$$cd(X) \leq \dim X.$$

See [2], exp. XIV, 3.2.

(11.3) The Base Change Theorem for Proper Morphisms

Given a cartesian diagram of morphisms of schemes:

$$\begin{array}{ccc} X' & \xrightarrow{f'} & Y' \\ {\scriptstyle g'}\downarrow & & \downarrow{\scriptstyle g} \\ X & \xrightarrow{f} & Y \end{array}$$

the base change morphism

$$g^*(R^q f_*(F)) \to R^q f'_*(g'^* F)$$

is defined for all $q \geq 0$ and all abelian sheaves F on $X_{ét}$ (cp. (1.4.7)). An interesting special case arises from the following situation:

Let $f : X \to Y$ be a morphism of schemes and let $g : P = spec(\Omega) \to Y$ be a geometric point of Y. Then we obtain the cartesian diagram

$$\begin{array}{ccc} X_P & \xrightarrow{f'} & P \\ {\scriptstyle g'}\downarrow & & \downarrow{\scriptstyle g} \\ X & \xrightarrow{f} & Y \end{array}$$

with the geometric fibre $X_P = X \times_Y P$ of $f : X \to Y$ in P. By (2.3) the functor $G \to G(P)$ induces an equivalence between the category of abelian sheaves on $P_{ét}$ and the category of abelian groups. If $h : Z \to P$ is an arbitrary morphism, and if G is an abelian sheaf on $Z_{ét}$, then the sheaf $R^q h_*(G)$ on $P_{ét}$ corresponds to the group $H^q(Z, G)$ (cp. the proof of (5.4)). Therefore the base change morphism of the above diagram is

$$(R^q f_*(F))_P \to H^q(X_P, F/X_P),$$

where F/X_P denotes the inverse image of F under $X_P \to X$.

The base change theorem for proper morphisms is the following:

(11.3.1) Theorem. *Let $f : X \to Y$ be a* **proper** *morphism of schemes and let F be an abelian torsion sheaf on $X_{ét}$. Then the base change morphism*

$$g^*(R^q f_*(F)) \to R^q f'_*(g'^* F)$$

is an isomorphism for all base extensions $g : Y' \to Y$ and all $q \geq 0$.

The proof of this theorem can be found in [2], exp. XII, XIII.

(11.3.2) Corollary. *Let $f : X \to Y$ be proper, F be a torsion sheaf on $X_{\text{ét}}$ and let $P \to Y$ be a geometric point of Y. Then we have*

$$(R^q f_*(F))_P \cong H^q(X_P, F/X_P)$$

for all $q \geq 0$.

(11.3.3) Corollary. *Let $f : X \to Y$ be a proper morphism of relative dimension $\leq n$. Then*

$$R^q f_*(F) = 0 \quad \text{for} \quad q > 2n$$

holds for all abelian torsion sheaves F on $X_{\text{ét}}$.

(11.3.4) Corollary. *Let $k \subset k'$ be separably closed fields, let X be a proper k-scheme and let $X' = X \times_k k'$. For all abelian torsion sheaves F on $X_{\text{ét}}$ we have*

$$H^q(X, F) \cong H^q(X', F')$$

for all $q \geq 0$. Here F' denotes the inverse image of F under $X' \to X$.

Besides the base change theorem for proper morphisms there is the following base change theorem for quasi-compact and quasi-separated morphisms:

(11.3.5) Theorem. *Let $f : X \to Y$ be a quasi-compact and quasi-separated morphism. Then the base change morphism*

$$g^*(R^q f_*(F)) \to R^q f'_*(g'^* F)$$

is an isomorphism for each **smooth** *base change $g : Y' \to Y$ and each ℓ-torsion sheaf F with ℓ invertible on Y.*

For the proof see [2], exp. XVI, 1.2.

(11.4) Finiteness Theorems

(11.4.1) Theorem. *Let $f : X \to Y$ be a proper finitely presented morphism of schemes. If F is a constructible abelian sheaf on $X_{\text{ét}}$, then the sheaves $R^q f_*(F)$ are constructible for all $q \geq 0$.*

The proof can be found in [2], exp. XIV, 1.

(11.4.2) Corollary. *Let X be a complete scheme over a separably closed field k. If F is a constructible abelian sheaf on $X_{\text{ét}}$, then the cohomology groups $H^q(X,F)$ are finite for all $q \geq 0$.*

(11.4.3) Theorem. *Let X be a smooth algebraic scheme over a separably closed field k. Let F be a locally constant finite abelian sheaf of order prime to $p = \text{char}(k)$. Then the groups $H^q(X,F)$ are finite for all $q \geq 0$.*

See [2], exp. XVI, 5.2.

Bibliography

Artin, M.
[1] Grothendieck Topologies, Mimeographed notes, Harvard University 1962

Artin, M., Grothendieck, A., Verdier J. L.
[2] Théorie des Topes et Cohomologie Étale des Schémas (SGA 4). Tome 1 - 3, Lecture Notes in Mathematics, vol. 269, 270, 305. Springer, Berlin Heidelberg New York 1972, 1973

Bourbaki, N.
[3] Commutative Algebra. Addison-Wesley, Reading 1973

Bucur, I., Deleanu, A.
[4] Introduction to the Theory of Categories and Functors. Wiley, London 1968

Cartan, H., Eilenberg, S.
[5] Homological Algebra, Princeton University Press, Princeton 1956

Coates, J., Lichtenbaum, S.
[6] On l-adic zeta functions. Ann. of Math. **98** (1973) 498 - 550

Demazure, M., Grothendieck, A.
[7] Schémas en groupes (SGA 3). Tome 1, Lecture Notes in Mathematics, vol. 151. Springer, Berlin Heidelberg New York 1970

Deligne, P.
[8] La Conjecture de Weil I. Publ. Math. IHES **43** (1976) 273 - 307

Deligne, P.
[9] La Conjecture de Weil II. Publ. Math. IHES **52** (1980) 137 - 252

Deligne, P., et al.
[10] Cohomologie étale (SGA $4\frac{1}{2}$). Lecture Notes in Mathematics, vol 569. Springer, Berlin Heidelberg New York 1977

Giraud, J.
[11] Cohomologie von Abéliennes. Springer, Berlin Heidelberg New York 1971

Giraud, J., Grothendieck, A., et al.
[12] Dix exposés sur la cohomologie des schémas (SGA 2). North Holland, Amsterdam 1968

Godement, R.
[13] Topologie Algébriques et Théorie des Faisceaux. Hermann, Paris 1958

Grothendieck, A.
[14] Sur quelques points d'algébre homologique. Tôhoku Math. J. **9** (1957), 119 - 221

Grothendieck, A.
[15] Fondaments de la Géométrie Algébriques. (Sém. Bourbaki 1957-62), Secrétariat Mathematique, Paris 1962

Grothendieck, A.
[16] Revêtements Étales et Groupe Fondamental (SGA 1). Lecture Notes in Mathematics, vol. 224. Springer, Berlin Heidelberg New York 1971

Grothendieck, A., Dieudonné, J.
Eléments de Géometrie Algébrique (EGA I - EGA IV)
[17] Le langage des schémas (EGA I). Springer, Berlin Heidelberg New York 1971

[18] Étude globale élémentaire de quelques classes de morphismes (EGA II). Publ. Math. IHES **8** (1961)

[19] Étude cohomologique des faisceaux cohérents (EGA III). Publ. Math. IHES **11** (1961), **17** (1963)

[20] Étude locale des schémas et des morphismes des schémas (EGA IV). Tome 1 - 4. Publ. Math. IHES **20** (1964), **24** (1965), **28** (1966), **32** (1967).

Hartshorne, R.
[21] Algebraic Geometry, Springer, Berlin Heidelberg New York 1977

Katz, N.
[22] An overview of Deligne's proof of the Riemann hypothesis for varieties over finite fields (Hilbert's problem 8). In: Mathematical developments arising from Hilbert Problems. AMS Proc.-Symp. Pure Math. **28** (1976), 275 - 306

Knus, M. A., Ojanguren, M.
[23] Théorie de la Descente et Algèbres d'Azumaya. Lecture Notes in Mathematics, vol. 389, Springer, Berlin Heidelberg New York 1974

Kurke, H., Pfister, G., Roczen, M.
[24] Henselsche Ringe und Algebraische Geometrie. Deutscher Verlag der Wissenschaften, Berlin 1975

Mazur, B.
[25] Notes on étale cohomology of number fields. Ann. Sci. Ecole Norm. Sup. **6** (1973), 521 - 556

Milne, J. S.
[26] Étale Cohomology. Princeton University Press, Princeton 1980

Mitchell, B.
[27] Theory of Categories. Academic Press, New York 1965

Raynaud, M.
[28] Anneaux locaux Henséliens. Lecture Notes in Mathematics, vol. 169. Springer, Berlin Heidelberg New York 1970

Šafarevič, I.
[29] Basic Algebraic Geometry. Springer, Berlin Heidelberg New York 1974

Schubert, H.
[30] Kategorien I. Heidelberger Taschenbücher 65. Springer, Berlin Heidelberg New York 1970

Serre, J.-P.
[31] Local Fields. Springer, Berlin Heidelberg New York 1979

Serre, J.-P.
[32] Cohomologie Galoisienne. Lecture Notes in Mathematics, vol. 5. Springer, Berlin Heidelberg New York 1964

Index

Universitext

Aksoy, A., Khamsi, M. A.: Methods in Fixed Point Theory
Aupetit, B.: A Primer on Spectral Theory
Bachem, A., Kern, W.: Linear Programming Duality
Benedetti, R., Petronio, C.: Lectures on Hyperbolic Geometry
Berger, M.: Geometry I
Berger, M.: Geometry II
Bliedtner, J., Hansen, W.: Potential Theory
Booss, B., Bleecker, D. D.: Topology and Analysis
Carleson, L., Gamelin, T.: Complex Dynamics
Carmo, M. P. do: Differential Forms and Applications
Cecil, T. E.: Lie Sphere Geometry: With Applications of Submanifolds
Chandrasekharan, K.: Classical Fourier Transforms
Charlap, L. S.: Bieberbach Groups and Flat Manifolds
Chern, S.: Complex Manifolds without Potential Theory
Chorin, A. J., Marsden, J. E.: Mathematical Introduction to Fluid Mechanics
Cohn, H.: A Classical Invitation to Algebraic Numbers and Class Fields
Curtis, M. L.: Abstract Linear Algebra
Curtis, M L.: Matrix Groups
Dalen, D. van: Logic and Structure
Das, A.: The Special Theory of Relativity: A Mathematical Exposition
Devlin, K. J.: Fundamentals of Contemporary Set Theory
DiBenedetto, E.: Degenerate Parabolic Equations
Dimca, A.: Singularities and Topology of Hypersurfaces
Edwards, R. E.: A Formal Background to Higher Mathematics I a, and I b
Edwards, R. E.: A Formal Background to Higher Mathematics II a, and II b
Emery, M.: Stochastic Calculus in Manifolds
Foulds, L. R.: Graph Theory Applications
Frauenthal, J. C.: Mathematical Modeling in Epidemiology
Fuks, D. B., Rokhlin, V. A.: Beginner's Course in Topology
Gallot, S., Hulin, D., Lafontaine, J.: Riemannian Geometry
Gardiner, C. F.: A First Course in Group Theory
Gårding, L., Tambour, T.: Algebra for Computer Science
Godbillon, C.: Dynamical Systems on Surfaces
Goldblatt, R.: Orthogonality and Spacetime Geometry
Gouvêa, F. Q.: *p*-Adic Numbers
Hahn, A. J.: Quadratic Algebras, Clifford Algebras, and Arithmetic Witt Groups
Hájek, P., Havránek, T.: Mechanizing Hypothesis Formation
Hlawka, E., Schoißengeier, J., Taschner, R.: Geometric and Analytic Number Theory
Holmgren, R. A.: A First Course in Discrete Dynamical Systems
Howe, R., Tan, E. Ch.: Non-Abelian Harmonic Analysis
Humi, M., Miller, W.: Second Course in Ordinary Differential Equations
for Scientists and Engineers
Hurwitz, A., Kritikos, N.: Lectures on Number Theory
Iversen, B.: Cohomology of Sheaves
Kelly, P., Matthews, G.: The Non-euclidean Hyperbolic Plane
Kempf, G.: Complex Abelian Varieties and Theta Functions

Universitext

Kloeden, P. E., Platen E., Schurz, H.: Numerical Solution of SDE Through Computer Experiments
Kostrikin, A. I.: Introduction to Algebra
Krasnoselskii, M. A., Pokrovskii, A. V.: Systems with Hysteresis
Luecking, D. H., Rubel, L. A.: Complex Analysis. A Functional Analysis Approach
Ma, Zhi-Ming, Roeckner, M.: Dirichlet Forms
Mac Lane, S., Moerdijk, I.: Sheaves in Geometry and Logic
Marcus, D. A.: Number Fields
McCarthy, P. J.: Introduction to Arithmetical Functions
Meyer, R. M.: Essential Mathematics for Applied Fields
Meyer-Nieberg, P.: Banach Lattices
Mines, R., Richman, F., Ruitenberg, W.: A Course in Constructive Algebra
Moise, E. E.: Introductory Problem Courses in Analysis and Topology
Montesinos, J. M.: Classical Tessellations and Three-Manifolds
Nikulin, V. V., Shafarevich, I. R.: Geometries and Groups
Oden, J. J., Reddy, J. N.: Variational Methods in Theoretical Mechanics
Øksendal, B.: Stochastic Differential Equations
Porter, J. R., Woods, R. G.: Extensions and Absolutes of Hausdorff Spaces
Rees, E. G.: Notes on Geometry
Reisel, R. B.: Elementary Theory of Metric Spaces
Rey, W. J. J.: Introduction to Robust and Quasi-Robust Statistical Methods
Rickart, C. E.: Natural Function Algebras
Rotman, J. J.: Galois Theory
Rybakowski, K. P.: The Homotopy Index and Partial Differential Equations
Samelson, H.: Notes on Lie Algebras
Schiff, J. L.: Normal Families of Analytic and Mesomorphic Functions
Shapiro, J. H.: Composition Operators and Classical Function Theory
Smith, K. T.: Power Series from a Computational Point of View
Smoryński, C.: Logical Number Theory I: An Introduction
Smoryński, C.: Self-Reference and Modal Logic
Stanišić, M. M.: The Mathematical Theory of Turbulence
Stillwell, J.: Geometry of Surfaces
Stroock, D. W.: An Introduction to the Theory of Large Deviations
Sunada, T.: The Fundamental Group and the Laplacian (to appear)
Sunder, V. S.: An Invitation to von Neumann Algebras
Tamme, G.: Introduction to Étale Cohomology
Tondeur, P.: Foliations on Riemannian Manifolds
Verhulst, F.: Nonlinear Differential Equations and Dynamical Systems
Zaanen, A. C.: Continuity, Integration and Fourier Theory